INTERPRETING EARTH HISTORY

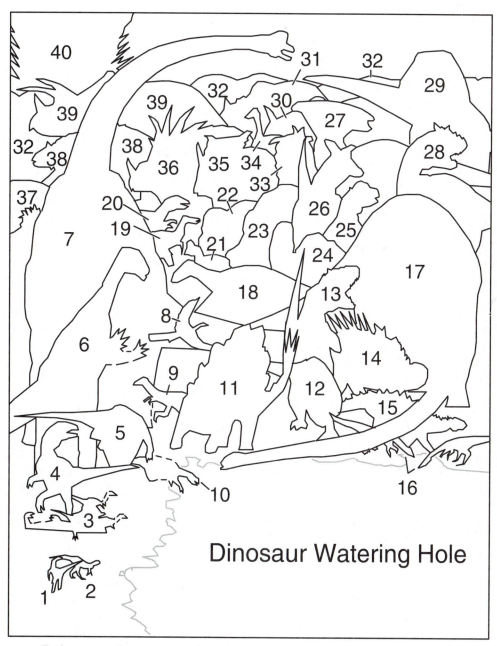

Dinosaur Watering Hole

Reference diagram for use in identifying the dinosaurs on the cover

INTERPRETING EARTH HISTORY
A MANUAL IN HISTORICAL GEOLOGY

Morris S. Petersen
Brigham Young University

J. Keith Rigby
Brigham Young University

Boston   Burr Ridge, IL   Dubuque, IA   Madison, WI   New York   San Francisco   St. Louis
Bangkok   Bogotá   Caracas   Lisbon   London   Madrid
Mexico City   Milan   New Delhi   Seoul   Singapore   Sydney   Taipei   Toronto

WCB/McGraw-Hill

A Division of The McGraw·Hill Companies

INTERPRETING EARTH HISTORY: A MANUAL IN HISTORICAL GEOLOGY,
SIXTH EDITION

1 2 3 4 5 6 7 8 9 0 QPD/QPD 9 3 2 1 0 9 8

ISBN 0-697-28290-2

Vice president and editorial director: *Kevin T. Kane*
Publisher: *Edward E. Bartell*
Sponsoring editor: *Robert Smith*
Developmental editor: *Lu Ann Weiss*
Marketing manager: *Lisa L. Gottschalk*
Project manager: *Donna Nemmers*
Production supervisor: *Deborah Donner*
Coordinator of freelance design: *Michelle D. Whitaker*
Photo research coordinator: *John C. Leland*
Art editor: *Jodi K. Banowetz*
Supplement coordinator: *Candy M. Kuster*
Compositor: *Shepherd, Inc.*
Typeface: *10/12 Times Roman*
Printer: *Quebecor Printing Book Group/Dubuque, IA*

Freelance interior and cover designer: *Cynthia Crampton*
Cover illustration: *Dinosaur watering hole by Dr. Stephen F. Greb, 1998*

www.mhhe.com

CONTENTS

PREFACE

Interpreting Earth History is intended for use in Historical Geology courses at the college and university level. As a supplement to lecture material, this manual allows the student to experience the challenges and satisfaction of using geologic data firsthand to derive meaningful interpretations. Topics are treated briefly; therefore, additional reading material or lecture information is essential to gain a working understanding of the concepts under consideration.

Of special concern to instructors are the number and type of exercises included in this manual. Some exercises are clearly meant to be done in a laboratory setting where rocks and fossil specimens are available (e.g., exercises 1, 9, 10, and 13). Other exercises can either be done in the lab or as homework assignments. No course in Historical Geology has time to include all 33 exercises. The intent of the manual is to provide a wide selection of exercises from which several can be chosen to fit the particular style of any Historical Geology course.

The sixth edition of *Interpreting Earth History* retains the organization and approximate size of the previous edition. Changes have been made to clarify text and student procedures, and to include new information. Illustrations and text have been changed to make the student assignments more explicit and straightforward. These modifications reflect numerous critiques by students and instructors who have found the book useful in the study of Historical Geology.

Over several years users of the manual have offered valuable advice and useful criticism. To all, I acknowledge my debt. I wish to extend a special thanks to those who have provided in-depth comments and suggestions in conjunction with the preparation of this edition.

M. S. P.

ONE
Principles

Review of Earth Materials

The fundamental building blocks of earth materials are atoms. Specific combinations of atoms build solid, crystalline, inorganic substances known as minerals. **Minerals** are defined as natural, inorganic substances of reasonably definite chemical composition and an orderly internal atomic arrangement, which produce certain specific physical characteristics (e.g., color, hardness, density, etc.).

Rocks are combinations of one or more minerals that are consolidated by any one of several natural Earth processes. Based on their origins, rocks are classified into three groups: (1) igneous, (2) sedimentary, and (3) metamorphic. The rock cycle, diagrammed in figure 1.1, illustrates the relationships among the three groups of rocks and the Earth processes involved in their formation.

Igneous Rocks

Igneous rocks are those rocks that formed through crystallization of melted rock material during cooling. They may form either on or below the Earth's surface. The molten source material is called **magma** if it is below the surface or **lava** if it is on the surface. Magma and lava are derived by internal heat from the Earth, which is concentrated along convergent plate margins (fig. 1.2), divergent plate margins (fig. 1.3), or at any one of several hot spots, or plumes. Igneous rocks are classified into groups based on their texture (size, shape, and boundary relations of individual grains) and mineral composition, as shown in figure 1.4.

The texture of igneous rocks, both intrusive and extrusive, is a mosaic of interlocking crystals (fig. 1.5). Intrusive igneous rocks are characterized by large, easily visible crystals produced by a slow cooling rate. Crystals of extrusive igneous rocks are small—essentially microscopic in size—because the lava cooled quickly.

The common types of igneous textures are as follows:

Coarse-Grained (phaneritic): Large crystals that are easily distinguishable with the unaided eye, typical of intrusive rocks.

Porphyritic: Coarse-grained or fine-grained with two sizes of crystals, one much larger than the other; the larger crystals are called phenocrysts.

Fine-Grained (aphanitic): Microscopic mineral grains, typical of extrusive rocks.

Vesicular: Fine-grained mass with numerous air holes formed by trapped gas within the lava; found exclusively in extrusive rocks.

Glassy: A noncrystalline solid, having the appearance of glass; found in extrusive rocks that have cooled very rapidly.

Fragmental: Composed of rock fragments of variable sizes and shapes called **tephra;** occurs exclusively in extrusive rocks.

Procedure

Complete the igneous rock worksheet (page 6) for each specimen of igneous rock provided by your instructor.

1. List the minerals that you can identify in each specimen.

2. Identify the proper texture for each specimen.

3. Name the rock type.

4. Briefly summarize the probable origin and history for each specimen.

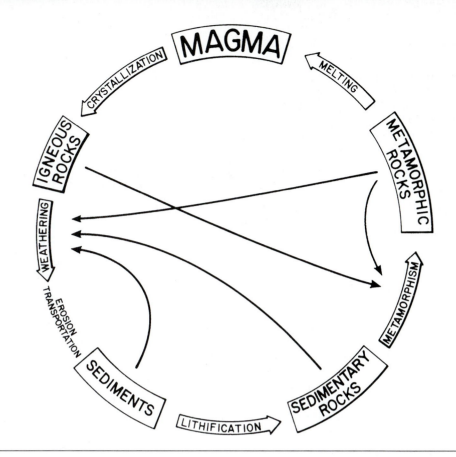

Figure 1.1 The rock cycle. The three main groups of rocks (igneous, sedimentary, and metamorphic) are produced by the operation of Earth processes. The relationship between the materials and the processes that form and modify them is shown.

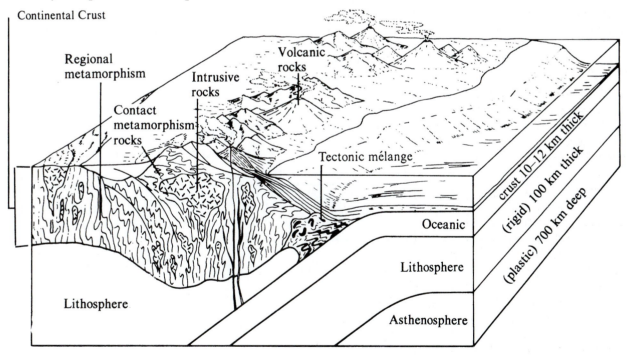

Figure 1.2 A block diagram showing the Earth's crust and upper mantle along a convergent plate margin. The genesis of igneous rocks, both intrusive and extrusive, is illustrated and labeled. The descending plate melts as it moves into deeper and hotter regions of the Earth's interior, providing the source of the magma and lava for the igneous rock bodies. Metamorphic rocks are formed by the heat and pressure associated with the magmatic masses and the compression generated by the converging plates.

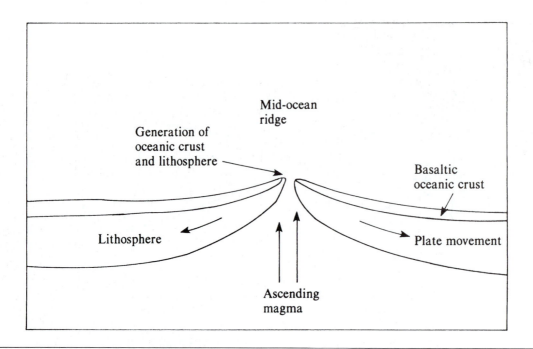

Figure 1.3 A cross section of a divergent plate margin at a mid-ocean ridge. Igneous rocks, which form the oceanic crust as well as the lithosphere, are formed here; the rocks then move away in both directions.

Texture		Color (Mineral Composition)		
		Light	Medium	Dark
Intrusive	Coarse-grained	Granite	Diorite	Gabbro
Extrusive	Fine-grained	Rhyolite	Andesite	Basalt
	Vesicular	Pumice		Scoria
	Glassy	Obsidian		
	Fragmental	Tuff and Breccia		

Figure 1.4 A classification of igneous rocks.

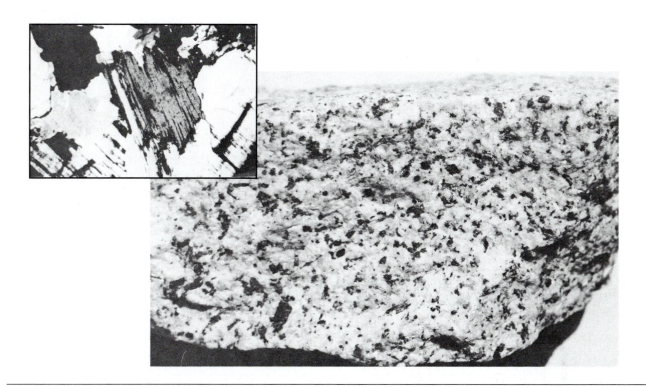

Figure 1.5 A hand specimen of granite, a coarse-grained igneous rock. The enlarged view of a thin section shows typical igneous texture, an interlocking crystalline pattern.

Sedimentary Rocks

Sedimentary rocks are the most common rocks on the Earth's surface. They are derived from mud, sand, gravel, and organic debris, all of which is collectively called sediment. Sediments accumulate under a variety of conditions giving rise to many kinds of sedimentary rocks (fig. 1.6). Sediments can accumulate as particulate material derived from eroded rocks, minerals, and organic skeletal debris. Such sedimentary rocks are called **clastic.** Clastic sedimentary rocks derived from organic debris, shells, plates, and the like, are further classified as bioclastics. Clastic particles range in size from clay particles to boulders. Other sediments form as a result of chemical precipitation, including both organic and inorganic processes. Another group of sedimentary rocks forms from the residue of the evaporation of mineral-rich water. These rocks are called evaporites. Coal, formed by accumulated, compacted, and altered plant remains, represents yet another variety within the spectrum of sedimentary rocks.

Accumulation sites of sediment include deserts, alluvial fans, glacial margins, streambeds, lakes, swamps, deltas, tidal flats, and a wide variety of near-shore and open marine environments (fig. 1.7). Compaction and cementation, or lithification, convert sediment into sedimentary rock (see fig. 1.1).

Sedimentary rocks are the most useful to the student of Earth history because they contain a reservoir of information about paleogeography, paleoecology, and extinct life. They also are the most widespread of all rocks that outcrop on the continents, amounting to 75% of the total. In a practical sense, sedimentary rocks are important because man's major energy sources (gas, oil, coal, and most uranium) are derived from these rocks.

Sedimentary rocks are variously classified on the basis of origin, grain size or texture, and chemical composition. A simple approach to the classification of the most common types of sedimentary rocks is shown in figure 1.8.

Igneous Rock Worksheet

Mineral Composition	Texture	Name	Origin and History

Igneous rock worksheet.

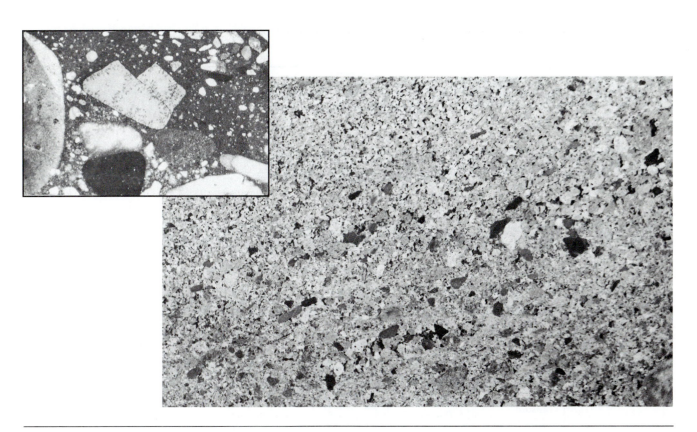

Figure 1.6 A hand specimen of coarse-grained sandstone. The enlarged view of a polished section shows typical sedimentary texture, a cemented mass of discrete sediment particles.

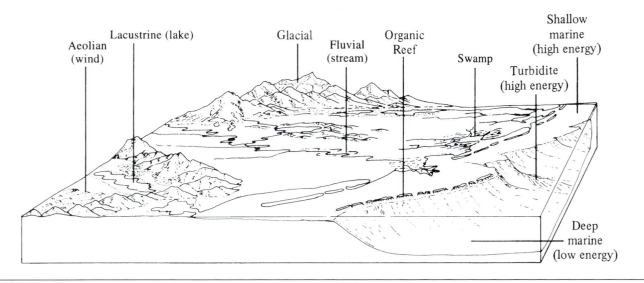

Figure 1.7 A composite drawing of nonmarine and marine sites for the origin of sedimentary rocks.

Origin	Sediment	Sedimentary Rock	Symbol	Characteristics
Terrigenous Clastics	Clay	Shale		Primarily clay and quartz grains of silt-size (<1/16 mm) or smaller having a thin platy structure.
	Sand	Sandstone	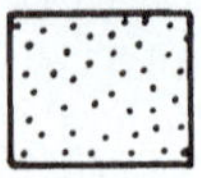	Sand-sized grains, (1/16 mm–2 mm) composed of quartz, feldspar and rock fragments cemented by silica, calcite, or clay.
	Gravel	Conglomerate		Rounded coarse-grained (>2 mm) rock particles usually cemented by silica or calcite.
Carbonates	Calcite $CaCO_3$	Limestone		Calcareous grains and skeletal fragments cemented with calcite, often containing fossils. Effervesces in dilute HCl.
	Dolomite $CaMg(CO_3)_2$	Dolomite		Dolomite grains commonly resulting from alteration of limestone. Effervesces only in powdered form.
Other	Plant remains	Coal		Lignite, bitumen, or anthracite, formed by the alteration of plant debris.
	Gypsum $CaSO_4 \cdot 2H_2O$	Gypsum	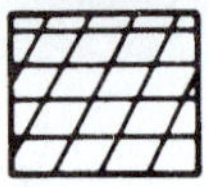	Occurs normally in sedimentary rocks as thin interbedded layers, formed by the evaporation of mineral-rich waters.
	Halite NaCl	Rock Salt	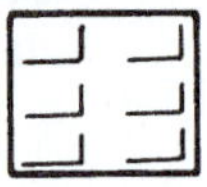	Accumulated by the evaporation of sea water.
	Silica SiO_2	Chert	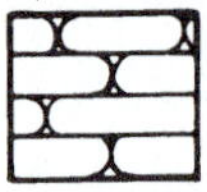	Dissolved from rocks by water, precipitated by both physical and biological means. Occurs in both fresh water and marine deposits.

Figure 1.8 Classification of sedimentary rocks, their sediment sources, and the symbols used for their representation in standard publications and throughout this manual.

Procedure

Complete the sedimentary rock worksheet (page 9) for each specimen of sedimentary rock provided by your instructor.

1. List the characteristics of each rock that relate to its classification and mode of origin.

2. Name each rock specimen.

3. What is the probable environment of deposition for each rock?

4. What can you say concerning the history of the rock?

Sedimentary Rock Worksheet

Characteristics	Name	Environment	Origin and History

Sedimentary rock worksheet.

Parent Rock	Metamorphic Rock	Characteristics
Limestone or Dolomite	Marble	Coarsely crystalline, commonly white, though variable in color, effervesces in dilute HCl.
Shale	Slate	Resembles shale, except much harder, cleavage plates form at angles to bedding of parent rock.
Quartzose Sandstone	Quartzite	Massive, hard, interlocking grains of quartz bound so tightly that fracturing will break through the individual grains of quartz.
Virtually Any Igneous and Sedimentary Rock	Schist	Mineral grains are elongated, producing a laminated appearance called foliation. Garnet and mica are common minerals.
Conglomerate	Metaconglomerate	Resembles conglomerate, except much harder, fractures break through the pebbles.
Impure Sedimentary Rocks and Granite	Gneiss	Mineral grains form sub-parallel light and dark bands.

Figure 1.9 Characteristics and parent materials of metamorphic rocks.

Metamorphic Rocks

Metamorphic rocks are those igneous and sedimentary rocks that have been altered by unusually high temperatures and pressures (fig. 1.9). Metamorphism changes the mineralogic and textural character of the original rock, producing a rock that is unique in texture, structure, and mineralogy. Some metamorphic rocks (gneiss and schist) display thin banding or layering in hand specimens (fig. 1.10). Metamorphic environments are produced by deep burial, tectonism, or igneous intrusion.

Common metamorphic textures include:

Foliated: Texture characterized by a planar orientation of the minerals present. Three types of foliation are classified in metamorphic rocks: schistosity, gneissic layering, and slaty cleavage.

Schistosity: Foliated texture resulting from the parallel arrangement of coarse-grained platy minerals, such as mica, chlorite, and talc.

Gneissic Layering: Foliated texture resulting from alternating layers of light (silicic) and dark (mafic) minerals.

Slaty Cleavage: Foliated texture resulting from the parallel orientation of microscopic platy minerals such as mica and chlorite. Slaty cleavage forms planes of weakness within a rock allowing it to split into slabs.

Phyllitic: A texture similar to slate but having a luster on the surface due to a concentration of mica minerals.

Nonfoliated: Texture characterized by a lack of planar orientation of the minerals present, producing a massive appearance. Quartzite and marble are examples of nonfoliated metamorphic rocks.

Procedure

Complete the metamorphic rock worksheet (page 12) for each specimen of metamorphic rock provided by your instructor.

1. List the characteristics of each metamorphic sample provided that relate to its classification.

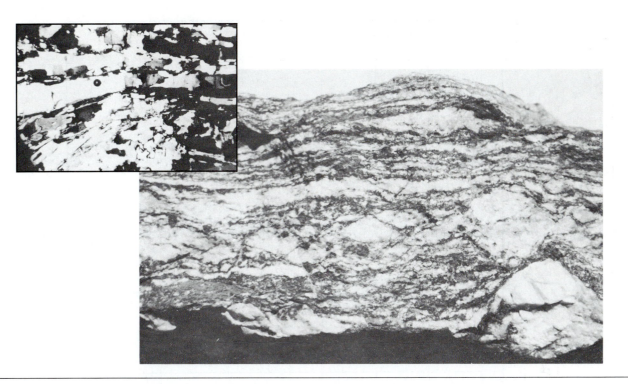

Figure 1.10 A hand specimen of gneiss, a banded metamorphic rock. The enlarged view of a thin section shows typical texture, a lineated pattern of mineral grains.

2. Name each rock specimen.

3. Identify the possible parent rocks for each specimen.

4. Briefly describe the probable origin and history for each specimen.

Metamorphic Rock Worksheet

Characteristics	Name	Parent Rock	Origin and History

Metamorphic rock worksheet.

The Principle of Uniformity

The **principle of uniformity** is basic and fundamental to all science. First described by James Hutton, a Scot, in 1785, it is a philosophy, or assumption, concerning nature that provides a way to scientifically deal with data about Earth history. The principle simply states that there is uniformity in natural law through time and space. Understanding natural laws and their effect on the processes operating on Earth today is a guide to past Earth events as well as future ones. It does not mean that today's world is an exact replica of either the past or future worlds, but only a sampling of each. Uniformity as a principle does not imply consistency in rate of change. This does not preclude catastrophic events as important in geologic history. Modern catastrophic events such as the 1980 eruption of Mount St. Helens and the 1964 Good Friday earthquake in Anchorage, Alaska, are examples of vigorous alteration of the Earth's features. Ancient catastrophic events are well documented in the geologic record, such as the erosion of the scablands of Washington State and the impact event thought by many to be the cause of extinction of many organisms at the close of the Cretaceous Period. To make that distinction clear, the word **actualism** has been suggested as an alternative to uniformitarianism. Geologic processes involved in the evolution of the Earth have always been, and will always be, governed by the same laws. Uniformity thus provides a basis for formulating models of the past in terms of our understanding of present geologic processes.

One hypothesis that derives from the uniformity principle is that geologic changes are, as a rule, slow; therefore, catastrophism, by itself, is not a very good model for understanding the evolution of the Earth.

Uniformitarianism can be applied, as an example, by examination of modern depositional environments and comparison with ancient sedimentary deposits. The environment under which sediments are deposited is highly varied with respect to physical and chemical conditions. Because of this variation, sedimentary rocks differ and reflect the unique conditions under which they were deposited.

The following outline describes various sedimentary environments and the resulting characteristics of each:

I. MARINE (deposited in oceans or marginal seas) (fig. 2.1)

  A. HIGH ENERGY (within wave base), OR ROUGH, SHALLOW-WATER ENVIRONMENTAL CHARACTERISTICS (littoral and shallow sublittoral):

   1. Deposition of medium to coarse sedimentary particles.
   2. Constant agitation and reworking of sedimentary material.
   3. Disarray and fracture of organic skeletal structures (shells).

   Resulting Sedimentary Rocks:

    a. Sandstone and conglomerate rock bodies.
    b. Moderately well-developed stratification in rocks.
    c. Fossils characteristically broken and randomly oriented.

  B. LOW ENERGY (below wave base) OR QUIET, DEEP-WATER ENVIRONMENTAL CHARACTERISTICS (sublittoral, bathyal, abyssal, and hadal):

   1. Quiet accumulation of fine sediments, rarely disturbed by water motions.
   2. Abundant benthonic organisms, both epifauna (surface) and infauna (sub-surface) dwellers.

   Resulting Sedimentary Rocks:

    a. Shale and siltstone rock bodies.
    b. Well-developed stratification, sometimes graded from coarse to fine upward in a sequence.
    c. Fossils are generally unbroken, and often composed of a large fraction of planktonic protozoa tests.

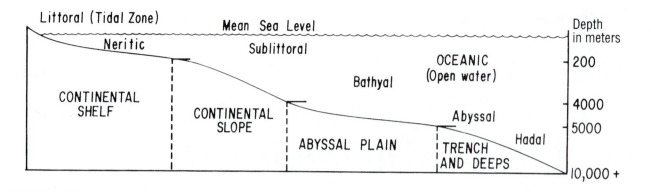

Figure 2.1   A classification of marine environments.

C. PELAGIC (open ocean deposits)
ENVIRONMENTAL CHARACTERISTICS:
1. Slow accumulation of skeletal debris, mainly from pteropods, diatoms, and radiolarians.
2. Slow accumulation of fine clay, marl, chalk, and volcanic particles.
3. All of the above material is slowly precipitated, or settled, from the open ocean water, far from any shore.
Resulting Sediment and Rock:
   a. Deposits of chalk, mud, limestone, dolomite, and chert.
   b. Carbon-rich, or organic-rich, sediment is common.
   c. Occasional small metallic or nonmetallic spherules of cosmic origin are added to the pelagic sediments.
   d. Potentially valuable deposits of ferromanganese nodules cover much of the deep North Pacific floor.

D. HIGH ENERGY DEEP WATER (turbidite)
ENVIRONMENTAL CHARACTERISTICS:
1. Extremely rapid transport of sediment moving from continents into the adjacent deep ocean environment.
2. Water and sediment move as undercurrents as a result of excess density driven by gravity.
3. Intermittent, catastrophic process that displaces large masses of sediment.
Resulting Sedimentary Rocks:
   a. Turbidite deposits show marked alternation in their sedimentary pattern.
   b. Sediment grades from coarse to fine upward in any particular layer.
   c. Some beds are laminar and thin-bedded; others are chaotic because of slumping after deposition.
   d. Deposits are generally composed of fine sand and silt.

E. ORGANIC REEF ENVIRONMENTAL
CHARACTERISTICS (fig. 2.2):
1. Vertical development of unbedded skeletal material, both living and dead, whose upper surface is above wave base (reef proper).
2. Seaward accumulation of eroded fragments of reef (fore reef).
3. Lagoon where fine calcareous sand may accumulate in flat-lying deposits; abundant organisms present (back reef).
Resulting Sedimentary Rocks:
   a. Massive, nonbedded calcareous mound or lens-shaped reef deposits consisting of intergrown fossils, including algae, sponges, corals, and other types.
   b. Fore-reef deposits of limestone, composed of reef-fragments in steeply inclined bedded strata.
   c. Back-reef lagoonal deposits of flat-lying and thin-bedded, highly fossiliferous limestone, but may grade to unfossiliferous dolomite or evaporite.

II. NONMARINE (deposits on land)
A. FLUVIAL (stream-deposited material)
ENVIRONMENTAL CHARACTERISTICS:
1. Transportation and deposition of poorly to moderately sorted, well-rounded rock particles, ranging in size from silt to boulders.
2. Deposition in irregular, elongated, and lenticular stream channels or valleys.
3. Terrestrial plants and animals living near streams.
Resulting Sedimentary Rocks:
   a. Deposits of conglomerate and sandstone rock bodies composed of well-rounded, poorly to moderately sorted rock particles.
   b. Poorly stratified deposit with lenticular beds and nearly planar, shallow cross-bedding.
   c. Fossils rare, consisting of land animals and plant remains.

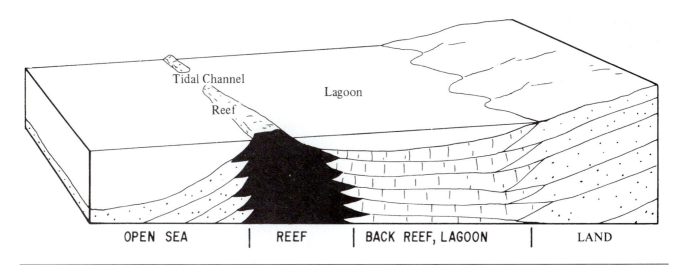

Figure 2.2   The organic reef environment.

B. LACUSTRINE (deposited in lakes)
    ENVIRONMENTAL CHARACTERISTICS:
    1. Quiet, slow accumulation of sediments with
        excellent sorting of particle sizes in interior;
        margin of lacustrine basin may have coarser
        beach zone.
    2. Freshwater animals and plants exclusively.
    Resulting Sedimentary Rocks:
        a. Well-sorted siltstone and shale rock bodies
            often showing highly regular continuous,
            alternating light and dark layers called
            varves, resulting from seasonal variation of
            depositional environment. Basin margin
            deposits may be coarse sandstone.
        b. Presence of fossils of freshwater organisms,
            plant and animal.
C. PALUDAL (swamp and marsh deposits)
    ENVIRONMENTAL CHARACTERISTICS:
    1. Slow, discontinuous accumulation of fine
        sedimentary material.
    2. Abundant plant life living and accumulating in
        the area.
    Resulting Sedimentary Rocks:
        a. Organic-rich shale and sandstone or coal
            deposits with thin stringers of siltstone and
            shale (partings).
        b. Plant fossils common in all stages of
            preservation.
D. EOLIAN (wind-blown deposits)
    ENVIRONMENTAL CHARACTERISTICS:
    1. Slow accumulation of extremely well sorted,
        fine-grained, often rounded sedimentary
        particles.

    2. Deposition of material as dunes with concave
        bedding on leeward side, changing direction of
        deposition with variation in wind direction.
    Resulting Sedimentary Rocks:
        a. Sandstone rocks composed of rounded and
            very well sorted particles of quartz or other
            resistant minerals. The surface of each well-
            rounded particle is often pitted or frosted.
        b. Well-developed, moderately high concave
            cross-beds up to several meters in length.
E. GLACIAL ENVIRONMENTAL
    CHARACTERISTICS:
    1. Massive erosion and transportation of large
        amounts of unsorted, angular sediments;
        particles ranging from clay to boulder or large
        block size within and along the margins of a
        glacier.
    2. Reworking of sedimentary particles by action of
        meltwater with deposition as stratified deposits
        in front of melting glacier.
    Resulting Sedimentary Rocks:
        a. Drift or tillite composed of unsorted
            individual angular particles, which may be
            striated on their surface because of the
            grating action of rocks against each other;
            deposited in clayey matrix.
        b. Mound or ridgelike deposits of unsorted
            particles of all sizes called moraines.
        c. Stratified deposits of fine silt and sand-sized
            material; sometimes reworked by wind and
            deposited as loess.
        d. Topography is also distinctive with
            numerous depositional and erosional features
            unique to glaciation.

Procedure

The following three pages contain photographs of modern environments in the left column and sedimentary rock types on the right (figs. 2.3, 2.4, and 2.5).

1. Using the principle of uniformity as a guide, identify the sedimentary rock types and match each with the appropriate modern environment as described in the outline above.

2. For each association, list the observations that were used in your analysis.

3. Of those observations listed above, which one characteristic gives the strongest evidence in support of your conclusion?

4. A. On figure 2.6, label the following beds on the photograph:
   a. Turbidite bed
   b. Slump bed
   c. Pelagic bed
   B. Upon what part of the modern ocean floor could sediment like this be accumulating? What data are available to verify your answer?

Low Energy Marine Environment, Grand Cayman Island,
B.W.I.

A. _____
(Height of the outcrop is approximately 20 meters.)

Paludal Environment, Everglades, Florida.

B. _____
(Note the pattern of bedding within the rock mass from right to
left.)

Organic Reef Environment, Raiatea, South Pacific.

C. _____
(Enlargement shows well-rounded rock particles approximately
6 cm in diameter.)

Figure 2.3

High Energy Marine Environment, North Carolina Coast. (Photo by H. J. Bissell.)

D. _____
(Enlargement shows accumulation of sediment in above area.)

Eolian Environment, Saudi Arabia. (Photo courtesy of Arabian American Oil Co.)

Coal

E. _____
(Coal seam is approximately 20 cm thick.)

Glacial Environment, Mendenhall Glacier, Juneau, Alaska.

F. _____
(Enlargement shows essentially entire fossil corals showing little, if any, evidence of transport.)

Figure 2.4

Fluvial Environment, Provo River, Utah.

G. _____

(The solitary corals are approximately 2 cm in diameter.)

Figure 2.5

Figure 2.6 Deep sea plain, pelagic, and turbidite beds in the western Pyrenees Mountains, Spain. For scale, the thickest light-colored bed is approximately 15 cm thick. Top of bedding is toward the left.

Subdivision of a Rock Column

As one can see when looking at the side of a mountain or the wall of a canyon, rocks, especially sedimentary rocks, typically show distinctive layering or stratification. This stratification is largely due to color and lithologic differences between adjacent beds, or groups of beds. This visual difference allows the classification and naming of rock units. In any normal stratigraphic sequence, rock masses can be classified into smaller and smaller more homogeneous units, and thicknesses of rock sections can be reduced to a series of manageable divisions, allowing effective study and description, as well as communication concerning them.

The basic unit of subdivision of a geologic sequence is a formation. A **formation** is a homogeneous rock unit, or an association of distinct interbedded rock units, which is separable from rock units above and below and can be shown on a geologic map of normal dimensions. A formation may consist almost entirely of cross-bedded sandstone, alternating sandstone and shale, or interbedded sandstone, shale, and limestone, as long as the unit is easily separable from rock units above and below.

Formations may be of any mappable thickness. In some areas it is useful to have formations only a few meters thick, whereas in other areas, because of the great thicknesses of the sedimentary column and its relatively uniform nature, formations may be several hundred meters thick. The accepted lower limit is based on local usefulness and whether the formation is sufficiently thick to appear on a map of a scale of approximately 1 inch per mile. Subdivision to units less than approximately 20 meters thick is not functional in most regions, although occasionally isolated distinctive key beds have been differentiated that are considerably thinner than 20 meters.

Formation boundaries are called **contacts** and are normally drawn at horizons showing marked lithologic change, such as where a limestone section changes to sandstone or shale, or where a sandstone changes to conglomerate. Contacts may also be placed at some arbitrary marker bed in a transitional lithologic series such as at a thin distinctive conglomerate in the middle of a sand and siltstone sequence, at a thick ash bed in the middle of a shale section, or at a distinctive limestone that separates sandy limestone from calcareous sandstone.

Formations are named from a locality such as a canyon, creek, or some prominent geographic feature where the rock sequence is particularly well exposed. This area is termed the **type locality** for the rock unit. The second part of the formation name is usually the dominant rock present in the formation such as shale, limestone, basalt, or granite. If the formation consists of a lithologic mixture, then the term *formation* is applied, as in Calvert Formation, Oquirrh Formation, or Morrison Formation, named after the town of Morrison, Colorado. These give some idea of the geographic occurrence, as well as the kinds of rocks, that are present in the unit.

A new formation is measured and described, the contacts are defined, and a specific section is designated for subsequent investigators as the best place to observe the formation and its relation to other formations. This locality is called the **type section** of the formation and it is usually in the same area as the type locality. If geographic names are limited and exposures are only locally suitable, it is possible to name a formation from one area and designate another exposure as the type section. For example, in Utah, the type locality of the Manning Canyon Shale is in Manning Canyon in the Oquirrh Range, but the type section is in Soldier Canyon, a few kilometers to the north, where the entire formation is better exposed.

Procedure

Part A

A somewhat generalized stratigraphic section of the rock column exposed in east-central Utah is presented in figure 3.1, utilizing the symbols for various rocks shown in figure 1.8. The somewhat irregular crossing dotted pattern in the lower central part of the column is cross-bedded sandstone.

The evenly bedded dots above represent a relatively uniformly bedded sandstone with some interbedded shale shown by the broken lines.

Gradations and variations in the basic rock patterns can be shown by combinations of symbols. A shaly limestone is shown as a dominant limestone pattern but with some broken lines. Siltstone is shown by an alternating dot and dash symbol, and sandstone that is slightly argillaceous is shown as a predominant dot pattern with only minor horizontal, broken lines. Sandy conglomerate is shown with a dotted pattern around the rounded symbols for pebbles. **Unconformities,** or surfaces of erosion, are shown by irregular wavy lines.

The relative resistance to erosion is shown by the right margin of the log. Cliff-forming units extend to the right; slope-forming units are shown by the indented line of the margin. In central and eastern Utah, limestone and some of the prominent sandstone formations are cliff-forming units. Most of the shale units and shaly sandstone beds recede to form slopes or valleys.

1. Based upon your analysis of the column, subdivide the rocks in figure 3.1 into formations, using the lithology and erosion characteristics as a guide. Mark your subdivisions or formations with heavy horizontal lines and define the contacts.

2. Using arbitrary localities, give a name to each of the formations, based on a geographic locality and the dominant lithology. Use the term *formation* where it seems appropriate because of nearly equally common interbedded lithologic types.

3. Give thicknesses of your designated formations by using the graphic scale, and then briefly describe the lithologic content of each formation.

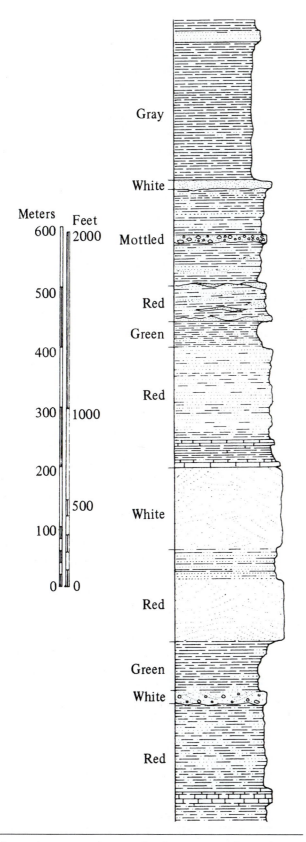

Figure 3.1 A sequence of sedimentary rocks exposed on the east flank of the San Rafael Swell in east-central Utah. Lines projecting to the left of the column show color breaks as noted.

4. Compare your classification with that of the actual subdivision supplied by your instructor. What effect has a regional knowledge of sequences in other nearby areas had on the choice for subdivision as compared to your interpretation?

Part B

1. On the geologic cross section shown in figure 3.2, repeat the process of subdivision into formations and into larger units. The diagram is a generalized cross section across the southeastern border of Wales and shows the relationships of major rock bodies in areas where much of the early subdivision of the geologic column was established. Because subdivision of the column is somewhat arbitrary, contacts are commonly placed at major changes in lithology or at major breaks in Earth history that record changes in environments or periods of erosion following folding and faulting. It was subdivision of rock sequences such as this, based on a unique fossil content, that resulted in development of the geologic column (fig. 3.3).

   Recent changes in the absolute (radiometric) dates for the major boundaries of the geologic column, shown in figure 3.3, are due to recent work by the International Stratigraphic Commision. Consisting of members from many nations, this body has, as its task , the following:

(1) defining each major stratigraphic boundary on specific fossil occurrences; (2) locating the most favorable location in the entire world to serve as the **stratotype,** or reference section, for each boundary, and (3) obtaining an accurate age for these boundaries with several methods of absolute dating. The result of this effort is a modification of boundary ages and time intervals between adjacent geologic periods. For example, moving the Cambrian–Precambrian boundary from 570 million years to 545 million years decreased the duration of the Cambrian Period from 65 million years to 50 million years. All major stratigraphic boundaries are being studied and altered, as appropriate, with new firm definitions, designated and marked stratogypes, and newly determined absolute ages.

2. What do the various unconformities show? What is their significance? Are they useful in establishing major subdivisions of the rock sequence?

3. List the sequence of events that produced the section shown in figure 3.2.

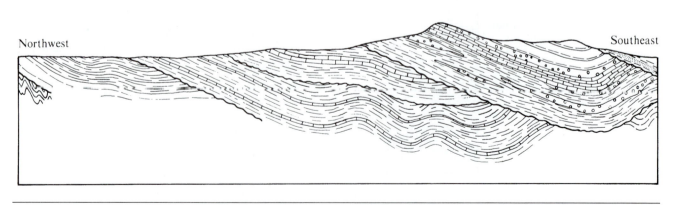

Northwest   Southeast

Figure 3.2 A restored cross section through the Paleozoic rocks along the southeastern border of Wales.

EON	ERA	Duration in millions of years	Millions of years ago
PHANEROZOIC	CENOZOIC	65	65
PHANEROZOIC	MESOZOIC	183	248
PHANEROZOIC	PALEOZOIC	297	545
PRECAMBRIAN / PROTEROZOIC	LATE	255	800
PRECAMBRIAN / PROTEROZOIC	MIDDLE	800	1600
PRECAMBRIAN / PROTEROZOIC	EARLY	900	2500
PRECAMBRIAN / ARCHEAN	LATE	500	3000
PRECAMBRIAN / ARCHEAN	MIDDLE	500	3500
PRECAMBRIAN / ARCHEAN	EARLY	500	4000

Era	Period		Epoch	Duration in millions of years	Millions of years ago
CENOZOIC	Quaternary		Pleistocene	1.7	1.8
CENOZOIC	Tertiary	Neo-gene	Pliocene	3.5	5.3
CENOZOIC	Tertiary	Neo-gene	Miocene	18.5	23.8
CENOZOIC	Tertiary	Paleogene	Oligocene	9.9	33.7
CENOZOIC	Tertiary	Paleogene	Eocene	21.1	54.8
CENOZOIC	Tertiary	Paleogene	Paleocene	10.2	65
MESOZOIC	Cretaceous			77	142
MESOZOIC	Jurassic			64	206
MESOZOIC	Triassic			42	248
PALEOZOIC	Permian			42	290
PALEOZOIC	Carboniferous	Pennsylvanian		33	323
PALEOZOIC	Carboniferous	Mississippian		31	354
PALEOZOIC	Devonian			63	417
PALEOZOIC	Silurian			26	443
PALEOZOIC	Ordovician			52	495
PALEOZOIC	Cambrian			50	545
PRECAMBRIAN					

Figure 3.3 Geologic time scale. Numbers are in millions of years.

Physical Correlation

To **correlate** is to make equal, or to establish comparable relationships. To correlate in a geologic sense is to establish either contemporaneity or continuity; that is, rock units may be equivalent in terms of time (**chronostratigraphy**) or in terms of being part of the same rock body (**physical correlation** or **lithostratigraphy**). Under unusual circumstances, rocks may be equivalent in both time and lithology but, generally speaking, one correlates rock units either in terms of time or lithology, not both.

Correlation is critical to an understanding of historical geology, because with it we can establish events in relationship to one another. Thus, one can gradually build a geologic column or a sequence of events and ultimately develop a geologic history. This exercise concerns correlation of rock units to establish lateral continuity or change.

Physical correlation denotes the use of physical criteria to accomplish the matching process. The goal of physical correlation is to establish the geographic extent, or continuity, of particular rock units. Physical correlation can be accomplished by any one or more of the following methods:

1. **Lateral continuity**
2. **Lithologic similarity**
3. **Sequence of beds**
4. **Geophysical characteristics** (seismic, electrical, sonic, radioactive, or magnetic properties)
5. **Sedimentary sequences**

We can demonstrate physical correlation by **lateral continuity** of single laminae or beds that can be traced from area to area. Single beds might be typified by ash falls, which geologically occur at a single instant over a wide area and hence offer ideal evidence of contemporaneity. The May 18, 1980, eruption of Mount St. Helens is a modern example of such an event. Volcanic ash was essentially instantaneously deposited over an enormous area extending from Washington to, after four days, the East Coast. We may correlate on **lithologic similarity** where unique kinds of rocks may be recognized in various areas and correlated. In the Colorado Plateau region of western North America, for example, the Navajo Sandstone is a distinctive, white, cross-bedded sandstone that retains much of its general characteristics over a vast area, even though the rocks above and below change somewhat in their composition.

We may correlate on the basis of a **sequence of beds** of distinctive lithologic units. Such correlations are more reliable than correlating on lithologic similarity alone. Statistically a sequence of beds is more reliable as an indicator of lateral continuity, particularly if the sequence exhibits a distinctive relationship clearly separating it from other rock units in the local area.

We may also correlate on various other **geophysical characteristics** of rock sequences such as the electrical, radioactive, sonic characteristics, and the like, of any given sequence. In many areas where subsurface geology is the primary exploration tool for petroleum, the rock sequence in most of the wells has been correlated by electrical, sonic, or radioactive properties. For example, a very weak current is generated where drilling fluid comes into contact with various formations. One can plot, as a curve or log against depth, the strength of this spontaneous-potential or self-generated current. This curve is distinctive in many instances. A curve also can be plotted of the depth against the resistivity of the various formations to an induced current. These curves can be compared and correlated even though the lithology of the sequence may be poorly known. A curve of natural radioactivity of the sequence is easily compiled. Such curves have proven very useful in many areas. Sonic logs, or curves that record variations in velocity of sound transmitted by individual rock layers, are also useful tools in correlation. Logs of each of these various physical properties, as well as others, are commonly composed of distinctive curves and may be correlated from region to region, based on unique points in the patterns.

Paleomagnetic effects offer still another method of physical correlation. The recent discovery of variations in the intensity or strength of the Earth's magnetic field over the past 160 million years has allowed determination of a series of major epochs of normal and reversed polarity that permit correlation on a worldwide scale. When magnetic, iron-bearing minerals crystallize or when they settle slowly out of the transporting medium, individual mineral grains align themselves with the existing magnetic field of the Earth. This produces a differential of direction of magnetism in various rocks, which is termed **remanent magnetism.** By measuring polarity, variations in intensities, and directions of inclination through a sequence of rocks or sediments, the sequences can be correlated on the basis of their paleomagnetic history. Remanent magnetic characteristics are particularly useful because they allow correlation between marine and nonmarine sediments, between volcanic rocks and sedimentary rocks, between very unlike sequences in widely separated areas, and, owing to their rapidity (approximately 5000 years), they provide a means of worldwide, nearly instantaneous, correlation.

Superimposed cyclic patterns in the stratigraphic record of marine rocks seem to reflect global climatic changes recorded as eustatic, or worldwide, synchronous, short-term changes in sea level, modified by longer-term tectonic changes, sediment supply, and climate. The interpretation and application of this concept leading to a new method of subdividing, correlating, and mapping sedimentary rocks is called **sequence stratigraphy.** Recognition of unconformity-bounded stratigraphic units called **sequences** has been in practice in geology for many years, largely beginning with the work of Prof. L.L. Sloss of Northwestern University in the late 1940s. A renewed interest in this method of interpreting marine strata evolved from **seismic stratigraphy** methods developed by P. R. Vail (a former graduate student at Northwestern) and others at Exxon Research during the 1970s.

Fundamental to the methods of sequence stratigraphy is the assumption that for all practical purposes, the seismic reflectors (fig. 4.1) used in identifying sequence units are time lines; thus, sequence stratigraphy is a chronostratigraphic method. A sequence represents one complete transgressive-regressive cycle of sea level, forming a sedimentary deposit representing a single genetic package bounded by unconformities, thus bounded by time lines. As sea level falls, the platform is eroded, and deposition takes place in the basin. When sea level rises abruptly, flooding the platform, onlap or transgression occurs as the water progressively deepens and sedimentation takes place more and more landward. At the high stand of flooding, there is practically no sedimentation accumulating in the basin forming a condensed interval of sedimentation. As sea level again falls, the sediment progrades toward the basin forming a downlap surface in the accumulated sediments.

A complex terminology, developed by Vail and his associates, must be mastered by those employing sequence stratigraphic methods. For the sake of introducing the concept of sequence stratigraphy, we will ignore the terminology and observe stratigraphic sequences to explore major sedimentary cycles.

Sedimentary cycles range in duration from approximately 20,000 years to 150 million years and occur as superimposed orders of cyclicity (fig. 4.2). The mechanism for changing sea level within the interval 20,000 to 500,000 years is thought by some geologists to derive from the Earth's orbital perturbations, the same forces often cited as the mechanisms that drove Pleistocene cyclic patterns.

Vail and his associates have published what have become known as "Vail Curves," which illustrate changes in sea level through time. Figure 4.3 is an example of a Vail Curve for part of the late Cenozoic (upper Eocene through Holocene).

Procedure

Part A

Four stratigraphic columns are illustrated in figure 4.4. One of these is the same stratigraphic column used in the exercise on subdivision of a geologic column. It is a column of rocks from Upper Permian beds to Upper Cretaceous beds exposed along the east side of the San Rafael Swell in eastern Utah. This column is used because it is relatively well known in geology. Column 1 is a stratigraphic section of beds of the same age exposed in the southern Wasatch Mountains in central Utah and from Westwater, near the Colorado–Utah line. For each of the four columns, locations are shown in the small index map.

1. Using the formation contacts supplied by your instructor for column 2 from the San Rafael district, correlate, using any means possible, to the three adjacent columns, showing lateral continuity of rock bodies or in some instances the disappearance of rock bodies between distinctive horizons. Unconformities are shown by irregular wavy lines. Color the various formations following the color designations along the margin of each column. After correlating the four sections, complete the following questions.

2. What specific methods did you use to correlate the following formations: the Navajo Sandstone, the Morrison Formation, the Chinle Shale, and the Kaibab Limestone?

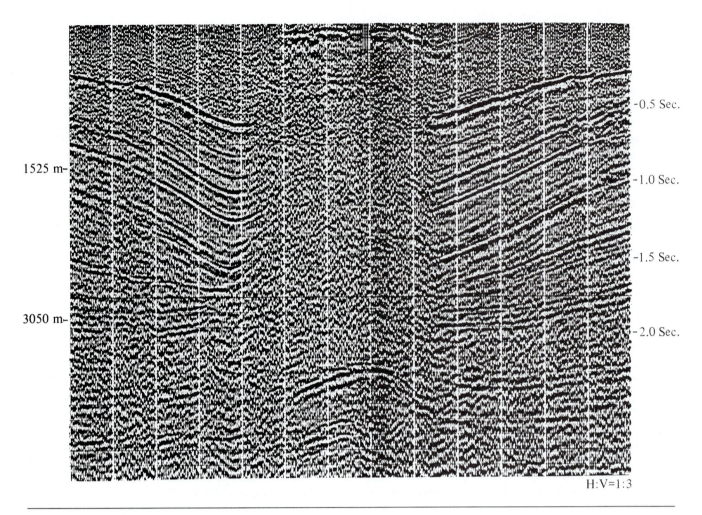

1525 m-

3050 m-

-0.5 Sec.

-1.0 Sec.

-1.5 Sec.

-2.0 Sec.

H:V=1:3

Figure 4.1   A reflection seismic profile from the Gulf
Coast.

SEQUENCE TERMINOLOGY	APPROXIMATE DURATION	EXAMPLES	AMPLITUDE (meters)	RISE/FALL RATE (cm/1000 yr)
1st Order	> 100 million years	Sequences of Sloss	——	< 1
2nd Order	10-100 million years	Supersequences	50-100	1-3
3rd Order	1-10 million years	Basic Sequence	50-100	1-10
4th Order	100,000 - 1 m.y.	——	1-150	40-500
5th Order	10,000 - 100,000 years	Parasequences	1-150	60-700

Figure 4.2   A classification of the units used in
sequence stratigraphic studies.

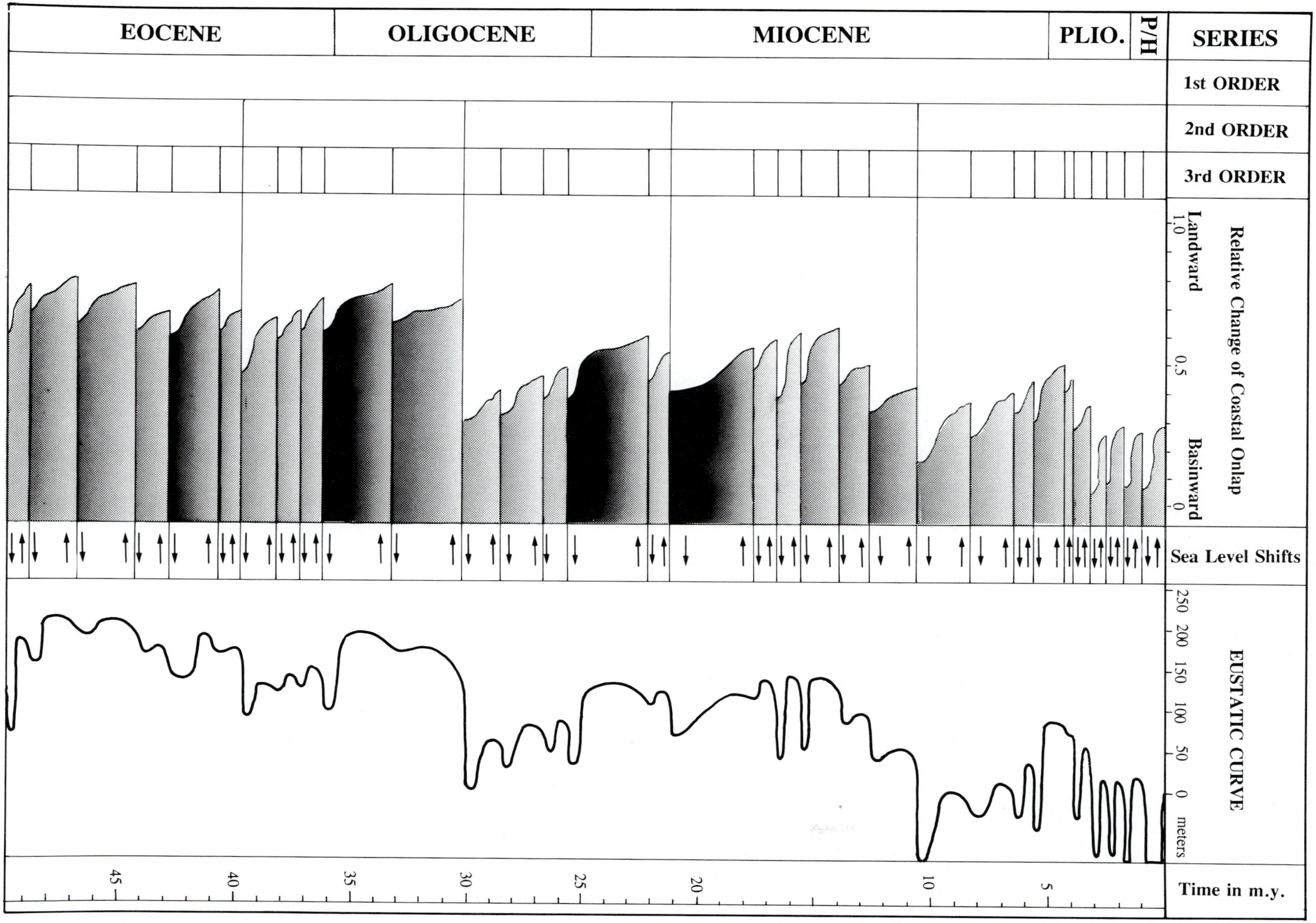

Figure 4.3 Eustatic cycle chart (Vail Curve) for the late Cenozoic (upper Eocene through Holocene). After Haq, B. U. et al., 1988. Mesozoic and Cenozoic Chronostratigraphy and Eustatic Cycles, in C. K. Wilgus, C. A. Ross, H. Posamentier and C. G. St. C. Kendall (eds.), Soc. Economic Paleon. and Mineral., Spec. Publ. 42, p.95.

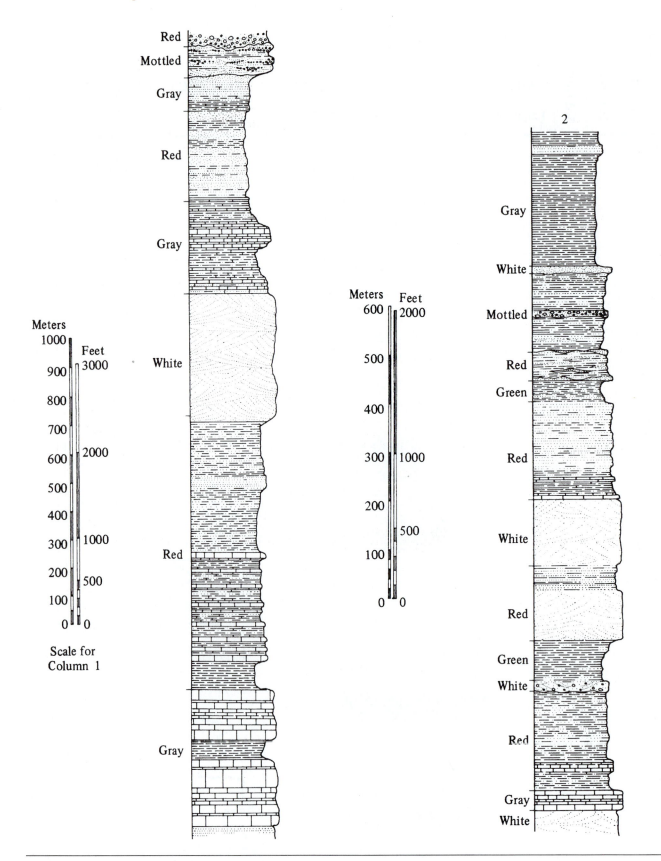

Figure 4.4 Four sequences of sedimentary rocks exposed in the Colorado Plateau and Wasatch Mountains in Utah.

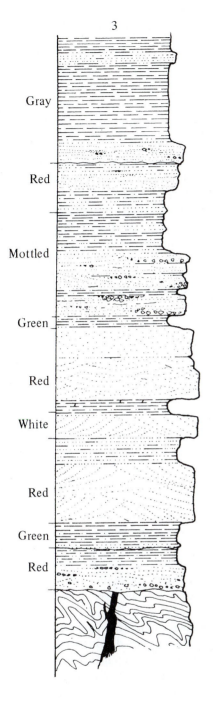

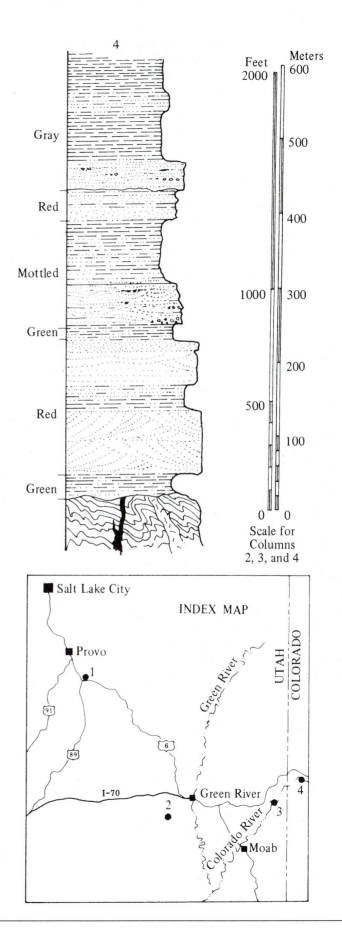

Figure 4.4 continued

3. Check your correlations with that of the instructor and obtain the proper names for use in the Wasatch Mountain, Westwater, and Green River columns.

4. How do you account for the differences in thickness and lithology of some units between these localities in the basal part of the column? How do you account for these differences in the middle part, above and below, the Navajo Sandstone?

Part B

Figure 4.5 illustrates three electric logs of wells drilled in southern Arkansas through Upper Cretaceous and Tertiary beds. The datum, or reference plane, is a level plane 1800 feet below sea level. The lithology and the names of the formations to be correlated are shown in the left-hand section. The two curves of electrical characteristics are the principal means of correlation. They can be correlated on the basis of very distinctive "kicks" or irregularities on the electric log. The curve of spontaneous-potential (S.P.) is to the left and the curve of resistivity (Res.) is on the right. Spontaneous-potential and resistivity curves for shale sections, in general, fall close to the median lines. Sandstone beds are more permeable and porous and contain freshwater, salt water, or natural hydrocarbons. Their curves swing away from the center lines on both resistivity and spontaneous-potential curves.

Marls or limestones appear much like slightly sandy shales on the spontaneous-potential curve, but swing away from the shale line on the resistivity curves. Footages below the derrick floor are shown by the tic marks and the depths along the left side of each of the logs.

1. Correlate the electric logs of the three wells using as many points as applicable.

2. These wells are approximately 10 km apart. Is this an area of strong folding or tilting of the sedimentary beds? How can you tell?

3. Is there evidence of thinning or wedging of beds at any particular stratigraphic level?

Part C

Figure 4.6 is a graph showing the history of polarity reversals in the intensity or strength of the Earth's magnetic field for approximately the last 5 million years. The bar graph shows 4 major epochs of reversed and normal polarity beginning with the Gilbert Reversal Epoch approximately 5 million years ago. During the following Gauss Normal Epoch the polarity was the same as that of modern times. The Matuyama Reversal Epoch was of long duration and preceded the present Brunhes Normal polarity.

During the 1960s, paleomagnetists unraveled a 4 million year record of a succession of reversals in the polarity of the Earth's magnetic field. The discovery of this phenomenon dates back to 1929 with the work of Professor Motonori Matuyama at the Kyoto Imperial University in Japan. With refinement in the potassium-argon dating method, precise dates could be obtained for relatively young (Pliocene and Pleistocene) volcanic rocks, allowing the further development of the geomagnetic-reversal time scale. More recent work has extended the polar reversal chronology record back 160 million years, using fossils recovered in deep-sea coves as the basis for the ages of the rock.

Three short-duration events interrupt the major oscillation of the epochs. The Mammoth Event took place approximately 3 million years ago, the Olduvai Event approximately 2 million years ago, and the Jaramillo Event approximately 1 million years ago. Utilizing the paleomagnetic history, a polar reversal chronology has been developed that can be utilized worldwide, even in sequences that greatly contrast in lithology.

The three stratigraphic columns in figure 4.7 represent three sequences that would be virtually impossible to correlate were it not for their magnetic character. The rate of sedimentation and the characteristics of the rock sequences are varied. Unconformities are shown in the stratigraphic columns by irregular wavy lines. Lithologic symbols are those used in previous exercises except for the very closely spaced vertical pattern, which is used to indicate basaltic volcanic rocks.

1. Using pencil lines, correlate the magnetic events and epochs in column 1, which is a log of a submarine core, with the magnetic patterns in columns 2 and 3, which are surface sections of a series of sediments and lava flows.

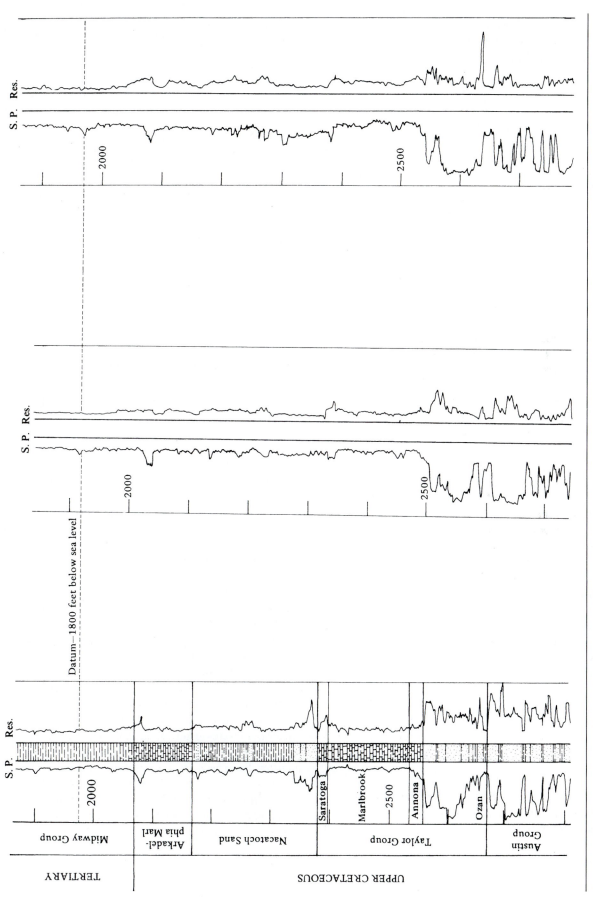

Figure 4.5 Electric logs of Upper Cretaceous and Tertiary rocks in southern Arkansas.

TIME IN M.Y.B.P (MA)	MAGNETIC POLARITY / ANOMALY NUMBERS	
0.00		
0.25	Brunhes (1)	Brunhes 0.72
0.50		
0.75		
1.00	Jaramillo	
1.25		0.72–2.47
1.50		
1.75	Olduvai	Matuyama
2.00	2 Reunion	
2.25		
2.50		
2.75	2A	2.47–3.40
3.00		
3.25		Gauss
3.50		
3.75		
4.00		
4.25	3	3.40–5.41
4.50		
4.75		Gilbert
5.00		
5.25		
5.50		

Figure 4.6 A graph of paleomagnetic polarity reversals for the past 5 million years. *Four* major epochs can be recognized along with *four* minor events.

2. What are the ages of the unconformities shown in columns 2 and 3?

3. What types of unconformities are illustrated in columns 2 and 3?

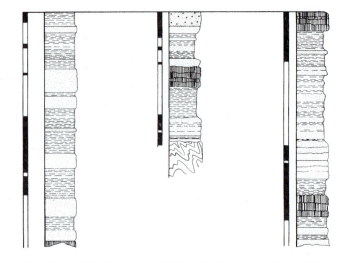

Figure 4.7 Three sequences of sedimentary and igneous rocks with patterns of paleomagnetic polarity illustrated as a bar graph. The dark portion of the bar graph indicates normal polarity, and the light indicates reversed polarity. Column 3 illustrates the effect of rapid deposition of rock strata. Columns 2 and 3 are incomplete with less than 4 million years of rock present.

Part D

Another method of physical correlation is related to the properties of transmittal and reflection of seismic or earthquake energy through the layered crust of the Earth. Early in the history of geophysics, it became apparent that energy pulses generated at the surface penetrated the upper crust with part of the energy being reflected back from horizons where markedly differing seismic properties were encountered. The energy appears much like light reflected through a series of stacked glass plates with some energy reflected and some transmitted through each of the plates (fig. 4.8). If sudden pulses of energy are generated by explosions of dynamite or gas or by vibrating a heavy weight on the Earth's surface, the energy is transmitted into the Earth's crust and part is reflected back at each reflecting horizon encountered at depth. This energy is picked up by a **seismometer,** an instrument that is sensitive to ground movement, and plotted as a record or curve of the motion of the Earth on a **seismogram.** Some horizons reflect more energy than others and produce larger motion in the seismometer. Such beds are most useful in seismic studies and appear in records as high-level reflecting horizons. If the elapsed time between initiation of the primary energy source and arrival back at the surface of the reflected

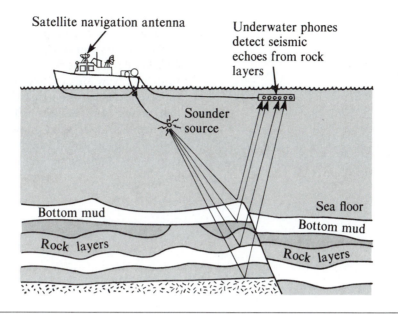

Figure 4.8 Seismic data collecting in offshore oil and gas exploration.

energy is plotted, the relative depths and velocity of the energy pulses through the rocks or sediments can be established. If the seismometer is moved, a series of seismograms are produced with distinctive horizons of peaks of high reflectance. This series can be arranged into a **profile** such as the one shown in figure 4.1. In figure 4.1, the horizons of high reflectance are shown as relatively light zones on the numerous seismic curves.

1. Correlate key reflecting horizons across the seismic profile by highlighting with a felt-tip pen or colored pencils. Show faults or folds where these are evidenced by offsets or flexures on the various reflecting horizons.

2. What methods of correlation did you use as you worked the problem? (Review the list of methods under **physical correlation** at the beginning of this exercise.)

3. What is the explanation for the poor definition of the reflecting surfaces in the center of the cross section?

4. Where is the most likely place to drill for oil in this section?

Part E

After studying figure 4.9, and comparing it with the photograph in figure 4.10, subdivide the Honaker Trail Formation into logical cycles of sedimentation (5th Order) by noting the shifts in sea level represented by the changes in types of sediments within the formation. Mark your cycles with brackets.

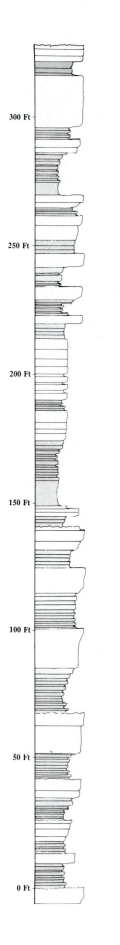

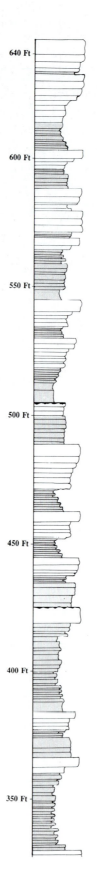

Figure 4.9 A stratigraphic section through the Middle Pennsylvanian Honaker Trail Formation in the Paradox Basin of the four corners area of Utah, Colorado, New Mexico, and Arizona. The shaded areas represent clastics; the nonshaded areas are carbonates. (After Goldhammer, R. K. et al., 1991. Sedimentary Modeling: Computer Simulations and Methods for Improved Parameter Definition, Kansas Geol. Surv., Bull 233, p. 369.)

Figure 4.10 Photograph of the Middle Pennsylvanian section exposed at Honaker Trail adjacent to the San Juan River near Mexican Hat, Utah. (Photo by Scott Ritter.)

Facies Relationships

The term **facies** is used in many ways in geologic literature. In a broad sense, it refers to any aspect of a rock, including appearance, composition, environment of formation, and any lateral changes, or variation, in these attributes over a geographic area, all of which reflect the environment under which the rock was originally deposited. You used facies interpretation in completing exercise 2, the exercise on uniformity. Facies is also used to denote essentially contemporaneous rocks of different lithology, or type, caused by environmental differences; for example, a rock body of limestone might be the time equivalent of an adjacent and interfingering formation of shale. Such an association is commonly developed in nearshore marine environments (figs. 5.1 and 5.2). Facies changes in sedimentary rocks are commonly the result of variations in sedimentary environments during deposition. If one were to look at the continental shelf off any of the present coastlines, one would see considerable variation in the kinds of sediments being deposited: coarse gravel or sand along the beach zone, grading offshore into silt and clay. In some areas, organisms produce carbonate masses, or reeflike structures. If these same sedimentary belts continue through time, a series of blocks of similar sediments would be deposited that might be classed as sedimentary facies, or as **lithofacies** in contrast with the depositional pattern associated with particular biologic groups, which are termed **biofacies.**

Biofacies or lithofacies developments are the result of sediments or organisms responding to a variety of factors in the environment, such as texture of substrate, chemistry in terms of the amount of oxygen and carbon dioxide available, salinity, water turbulence, and turbidity (the amount of suspended sediment). These as well as other factors influence the characteristics of the sedimentary and biologic association that we see in the geologic record as facies of either sedimentary rocks or biologic associations.

Early geologists who studied the rocks of Europe and North America thought primarily in terms of uniform layers of rocks that extended virtually worldwide and that were deposited from a primeval sea. They had complex interpretations to explain areas where such supposed worldwide uniform patterns were not consistent. Near the turn of the century, lateral variations in contemporaneous strata were recognized, and a new study of facies relationships began.

If all aspects of an environment remained constant during some period in geologic time, a series of vertical belts would develop, such as illustrated in figure 5.1. Depth of water, composition and texture of sediments, rate of sedimentation, rate of subsidence, and other factors in this example have been held constant, building a series of sedimentary rocks with relatively constant patterns. Such constancy in the geologic record is rare, however, and generally the environment varies, resulting in a lateral shift in the kinds of rocks that are produced as illustrated in figure 5.2. In this relationship, the facies belts, or areas of sedimentation of various types of sedimentary rocks, have migrated toward the left through time, as a result of increase in the coarseness of the incoming sediment and of the amount of material being transported into the sedimentary basin. The rate of sedimentation was more rapid than the rate of subsidence, resulting in filling of the sedimentary basin on one side and ultimately crowding the shoreline to the left, expanding the landmass at the expense of the area covered by the sea. The thickness of sediment deposited during any one time interval, as shown in figures 5.1 and 5.2, remained relatively constant. There was no development of a thick wedge of clastic sediment toward the source area. If the rate of sedimentation were influenced by rapid subsidence in the area of coarse sediments, a marked thickening of the sediments deposited during any time interval should result, such as shown in figure 5.3. In the region to the right, the rate of sedimentation was more rapid than to the left and a thicker sequence of sediments accumulated. This is shown by a divergence of the time lines. In facies studies, therefore, one can learn something concerning the relative rates of sedimentation, direction of transport, and capacity of the transporting medium.

Figure 5.4 is a restored cross section of Cambrian rocks visible in walls of the Grand Canyon and shows an intertonguing relationship of dolomite and limestone in the west (left) with an eastern belt of shale. Time lines, established by the occurrence of distinctive fossil faunas, are shown as heavily dashed, approximately horizontal lines. The locations of measured stratigraphic sequences are

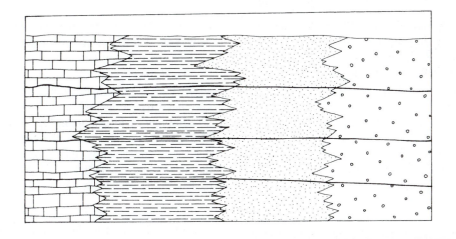

Figure 5.1   Schematic cross section showing facies relationships under static conditions of subsidence and sediment supply in a marine environment. The dark horizontal lines delineate synchronous units. Sediment was derived from the area to the right of the depositional site illustrated. The primary cause of sorting, as determined by coarser grain size, is water depth, or distance from shore.

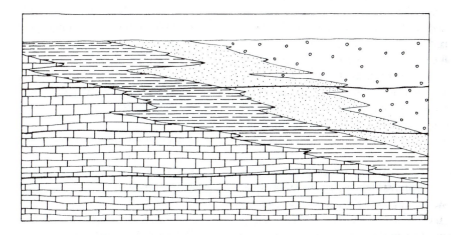

Figure 5.2   Schematic cross section showing facies variations produced by fluctuations in the amount and kind of sediment deposited under uniform conditions of subsidence. In figure 5.1, the sediment source area was to the right of this depositional site. Variation seen here would probably be caused by an uplift of the source area with an attendant increase in particle size and quantity of transported sediment.

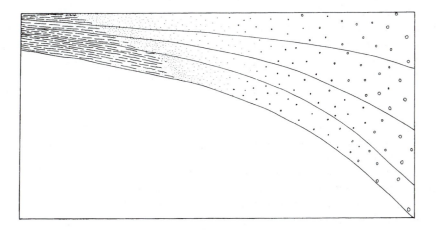

Figure 5.3   Schematic cross section showing facies relationships under conditions of varying subsidence and sediment supply. The curved lines represent time lines whose positions were determined by fossils. Source area for these sediments is to the right. Variation shown is due to subsidence of the depositional site near the source area.

shown by the heavily dashed, vertical lines. In the central part of the cross section, limestones can be seen grading laterally into thin tongues of dolomite that interfinger with shales in the upper part of the Bright Angel Shale. Intertonguing relationships such as these suggest very strongly that the limestones are time equivalents of the thin dolomite beds, and that the dolomite beds are the lateral temporal equivalents of the shale.

To the east, beyond the limit of the cross section, even these intermediate and upper beds grade into sandstone. The pattern is much like that shown in the lower part of the measured sections to the west, where Bright Angel Shale is demonstrated to be contemporaneous, or to have been deposited at the same time, as the beds of the Tapeats Sandstone.

In the exercise on physical correlation, it was emphasized that we correlate either on the basis of equivalence in terms of time or in terms of continuous rock bodies. In figure 5.4, the term Tapeats Sandstone is applied to all of the sandstone at the base of the Cambrian section, although beds in the eastern part of the exposure are younger than those in the west. On the basis of lateral continuity, stratigraphic position, and lithology, the Tapeats Sandstone beds in the west are correlated to the east as part of a continuous sandstone body, even though of slightly differing ages. One can speak of the sandstone facies at the base of the Cambrian rocks in contrast to the shale facies that overlies it, and in turn to the dolomite and limestone facies represented in the younger rocks.

To some extent, the Cambrian rocks demonstrate **Walther's law** or **principle,** which is: Those facies that occur adjacent to each other at the present time can be superimposed. For example, the Bright Angel Shale grades laterally into the Tapeats Sandstone and also overlies it. Similarly, in the central part of the section, dolomite and shale units are interbedded and grade laterally into one another. Although not infallible, the general observation holds that various kinds of rocks that are superjacent to one another in a stratigraphic sequence also grade laterally into one another along the outcrop band unless it is interrupted by some external event. This relationship can be seen in shale, sandstone, and conglomerate sequences, as well as in limestone, dolomite, and evaporite sequences.

Procedure

Part A

The first part of the exercise utilizes the Devonian rocks in New York and Pennsylvania. These are the rocks that formed the basis for the acceptance of the facies concept in North America. Fifteen somewhat generalized stratigraphic sections, which were measured through the Devonian rocks at localities approximately 32 km apart, are plotted as logs in figure 5.5. Section 1 is toward the west and section 15 is toward the east in a traverse that lies generally along the New York–Pennsylvania border. Symbols of the lithology are those used in previous exercises.

Total thickness of the preserved sections are shown to scale. Various time lines, which have been identified and correlated by the use of fossils, are shown by a series of dots through each column and are marked by small letters. The various time horizons or levels of contemporaneous deposition are shown by the same letters. For example, all the rocks immediately below the dotted line marked "a" in each of the 15 sections were deposited contemporaneously.

1. Construct a restored section for these Devonian rocks (fig. 5.5) similar to the example of Cambrian rocks in figure 5.4. Detach both pages of figure 5.5. Placing them side by side the long way, tape the two pages together and proceed. With lines and symbols, interconnect various lithologic units and show the facies relationships of the relatively coarse-grained rocks in the east to the fine-grained rocks in the west.

2. Does Walther's law apply to these rocks? Are there exceptions?

3. Are all of the conglomerates the same age?

4. What trend is visible in the sandstone beds as they are traced from east to west?

5. Why do shale beds thin as traced from west to east?

6. What happens to the sandstone that occurs near the base of sections 12, 13, 14, and 15?

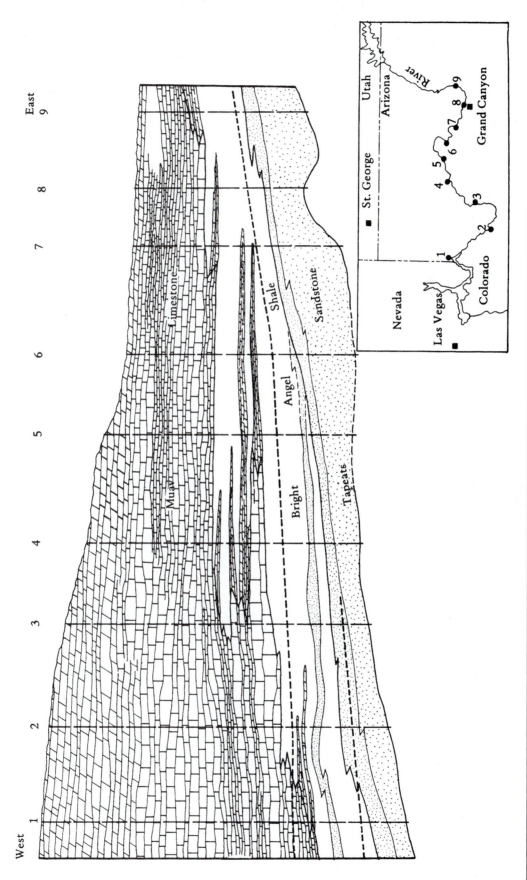

Figure 5.4 Restored cross section of Cambrian rocks in the Grand Canyon area of northern Arizona, from Lake Mead on the west eastward to the junction of the Little Colorado and Colorado rivers. (Courtesy Carnegie Institute of Washington, publication 563, October 1945 by E. McKee.) The horizontal distance from section 1 to 9 is approximately 220 km. The thickness of section 9 is approximately 600 meters.

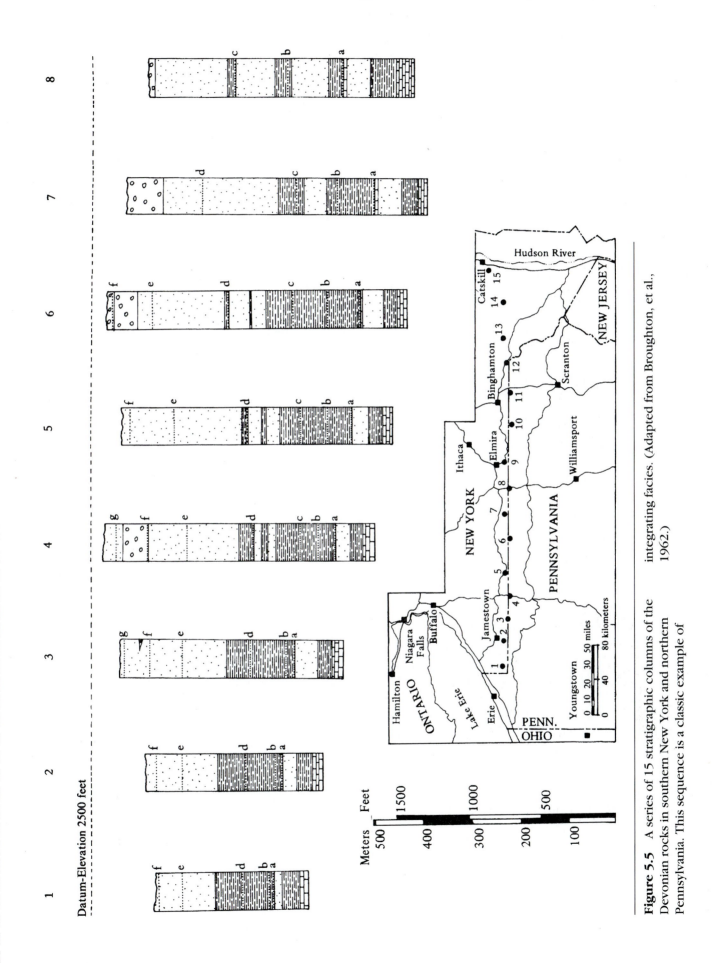

Figure 5.5 A series of 15 stratigraphic columns of the Devonian rocks in southern New York and northern Pennsylvania. This sequence is a classic example of integrating facies. (Adapted from Broughton, et al., 1962.)

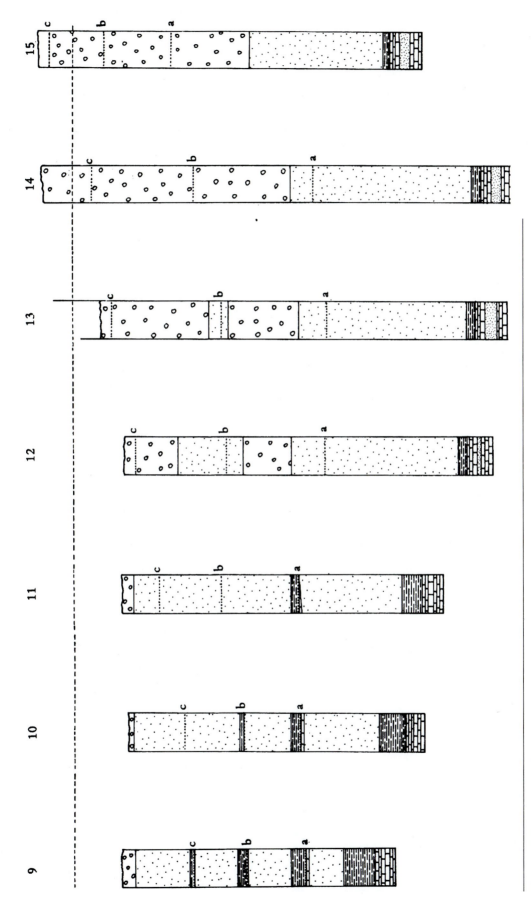

Figure 5.5 continued

7. What is suggested by the small lens of conglomerate near the top of section 3 in the time interval between "f" and "g"?

8. From which direction were the sediments transported?

9. By what media (e.g., wind, glaciers, streams, marine currents) were the sediments deposited?

Part B

Ten stratigraphic columns shown in figure 5.6 document the facies relationships through the classic Captain Reef in the Guadalupe Mountains of Texas and New Mexico, one of the major carbonate reef masses in North America (see fig. 2.2). This series of stratigraphic sections is oriented approximately northwest to southeast. Bedded limestone is shown with the normal bricklike symbol, but massive limestone is shown with an open discontinuous pattern. Gypsum, or evaporites, are shown with a close-spaced diagonal grid in sections at both the eastern and western margins.

1. Construct a restored cross section in the same manner as figure 5.5. The lines that connect adjacent stratigraphic sections are time lines and are labeled with letters (a–f) like those in the Devonian diagram of part A.
2. Are the massive reef limestones of section 3 the same age as the massive reef limestones of section 7? Are those of section 3 contemporaneous to those of the upper part of section 6?

3. What is the age of the massive dolomite in section 2 in relationship to the limestone beds of the Cherry Canyon Formation?

4. What is the age of the thin gypsum bed at the top of section 1 in relationship to the rocks in sections 8 and 9?

5. What is the direction of growth through time of the reef mass?

6. It is generally felt that the Cherry Canyon Formation contact with the top of the Brushy Canyon Formation remained essentially horizontal during deposition of these Permian reefs and that the top of the reef or the top of the massive dolomite was near sea level. In what depth of water was the gypsum in the top of section 1 deposited? What was the water depth during the deposition of the gypsum in sections 9 and 10?

7. From the diagram, which sequence would be termed *back reef* and which sequence would be termed *basin* or *fore-reef facies?*

Part C

The stratigraphic profile shown in figure 5.7 is an east-west series of stratigraphic sections through the Upper Cretaceous rocks of eastern Utah. The profile shows relationships of the clastic wedge associated with a pulse of the Sevier Orogeny in western North America.

1. Construct a restored cross section of these Cretaceous rocks using the same techniques as applied to the Devonian and Permian rocks in parts A and B. Use the same method previously used.
2. What is the direction of transport of the sediments?

47

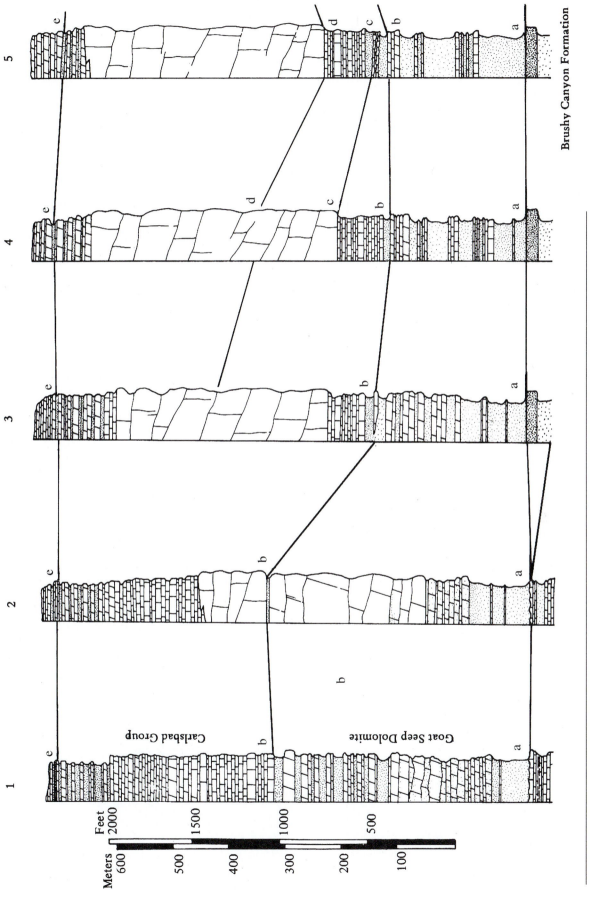

Figure 5.6 A series of 10 stratigraphic columns through the famed reef complex of the Guadalupe Mountains of western Texas and southern New Mexico.

(Adapted from King, P. B. 1948.) Compare with figure 2.2. Horizontal distance from section 1 to 10 is approximately 8 km.

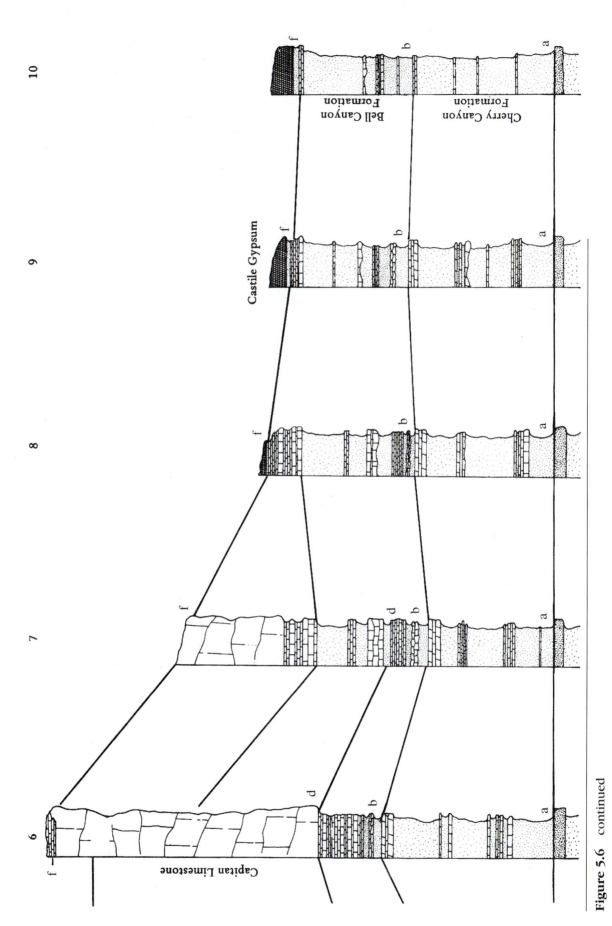

Figure 5.6 continued

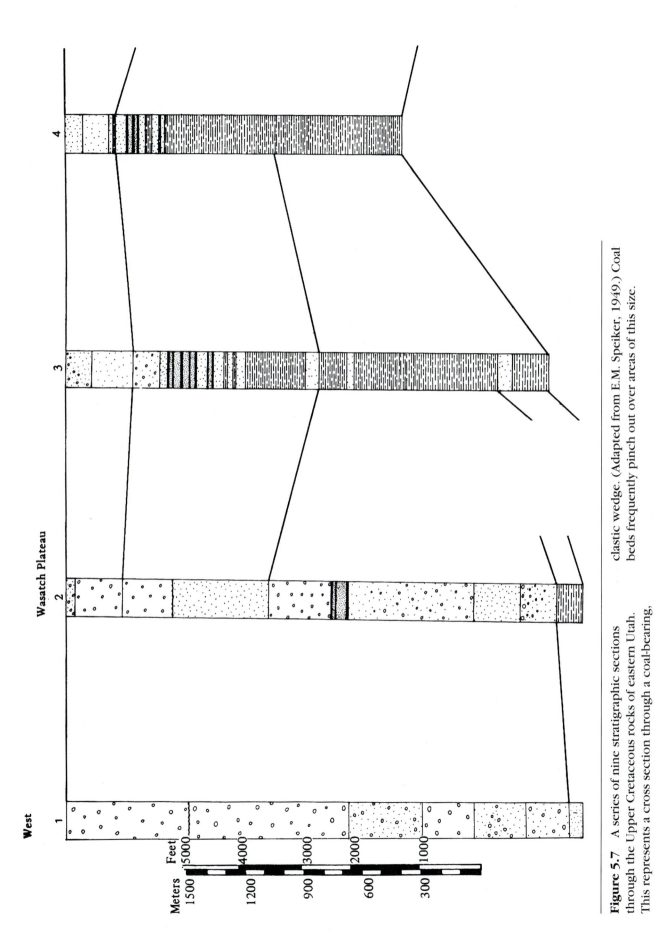

Figure 5.7 A series of nine stratigraphic sections through the Upper Cretaceous rocks of eastern Utah. This represents a cross section through a coal-bearing, clastic wedge. (Adapted from E.M. Speiker, 1949.) Coal beds frequently pinch out over areas of this size.

53

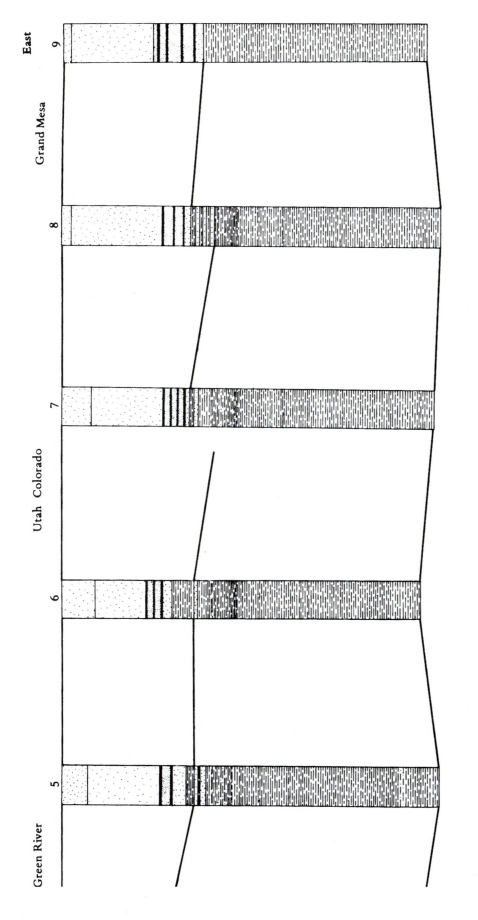

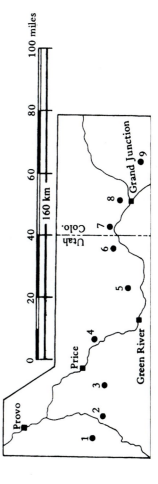

Figure 5.7 continued

3. What is the relative rate of subsidence compared to the rate of sedimentation? What is your evidence?

4. What is the environment of coal deposits?

5. Is the depositional pattern transgressive or regressive?

6. When was the most evident pulse of uplift in the source area during this part of the Sevier Orogeny? What are the evidences?

Part D

If one can document the kind of sediments deposited at any one "brief" interval in time over a wide area, one can construct lithofacies maps to show patterns of sedimentation. Figure 5.8 is a map of Upper Cretaceous deposits in the Rocky Mountains. The small circles are points where data are available, and lithology at each is shown by the following series of abbreviations: congl. for conglomerate, ss for sandstone, sh for shale, and m for marl or chalk. The line of zero thickness is shown along the west.

1. Construct a lithofacies map by drawing lines separating the various rock types that can be recognized.

2. Do the facies belts parallel the zero thickness line?

3. What is the direction of transport of the sediments? What is the probable source area for the coarse sediments along the western border of the map?

4. Were the Colorado Rocky Mountains present during deposition of these sedimentary rocks? What is the evidence?

5. Where would you expect the greatest thickness of sediments to have accumulated? Why?

Part E

Figure 5.9 is an isopach map, which is a map that illustrates the thickness of a specific rock sequence. The contour lines connect points of equal thickness over the area shown. This map indicates the thickness of the Upper Cretaceous rocks in the middle and southern Rocky Mountain area. The line of zero thickness is along the western margin. The area is the same as that of figure 5.8. Draw the top of the section level, and let the base vary.

1. Construct a restored cross section along the line A–A'. This line extends from Provo, Utah, to Denver, Colorado. Use the lithology shown on figure 5.8. Assume that the full thickness of rock for any locality is similar to that shown on figure 5.8. Some integration of rock types will be necessary to draw a realistic cross section (see figs. 5.5, 5.6, and 5.7).

2. What can you conclude as to the topography of western Utah during Upper Cretaceous time?

3. What orogeny is responsible for this sedimentary sequence?

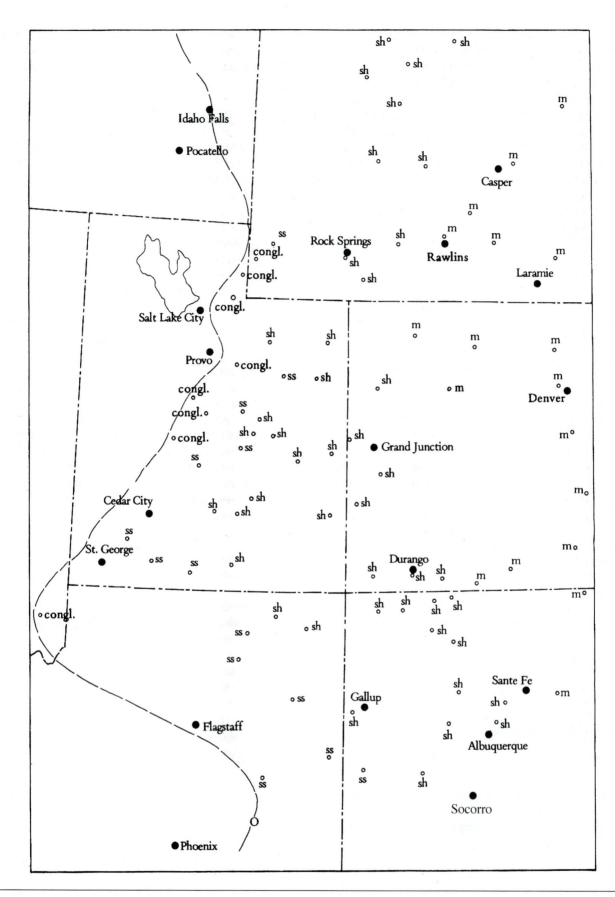

Figure 5.8 A map showing distribution of rock types of part of the Upper Cretaceous clastic wedge in the Rocky Mountains. This is an area between the stable area of the continent and the western mobile belt. Abbreviations: congl. = conglomerate, ss = sandstone, sh = shale, m = marl or chalk.

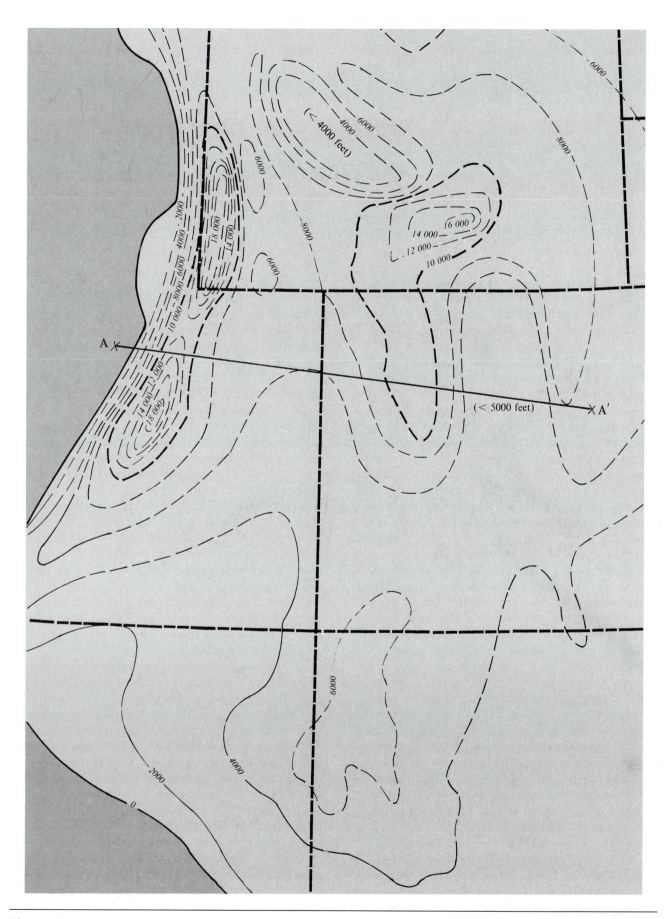

Figure 5.9 An isopach map of Upper Cretaceous strata in the middle and southern Rocky Mountain area.

The contours are solid where data are certain, dashed where inferred. Contour interval is 2000 feet.

Relative Dating—
Sequences of Events

The immensity of time is one of the unique features of the science of geology. In this exercise, however, we will be concerned only with the relative sequence of events; that is, event A preceded event B, or geologic feature A is older than geologic feature B but younger than geologic feature C. Such geologic dating, where one event is established as older or younger in relationship to another, is referred to as relative dating.

Relative geologic ages are established primarily with three fundamental concepts. First, sedimentary rocks, in the main, were deposited originally horizontally. Any marked variation from the horizontal or the bedding indicates some movement of the Earth's crust. The original horizontal attitude of most sedimentary units has been described in what is formally called the **principle of original horizontality.**

Second, those rocks that are highest in a normal undisturbed stratigraphic sequence are youngest, or conversely, those that are lowest in the undisturbed sequence were deposited first and are oldest. The second major principle in terms of relative dating has been formulated as the **principle of superposition.** In many canyon walls, for example in the Grand Canyon, rocks along the canyon rim were deposited over rock layers or formations exposed lower on the canyon walls. Thus, those beds along the canyon rim are younger than those exposed in the inner gorge and in the lower part of the Grand Canyon.

Third, geologic structures or rock bodies that crosscut other bodies or structures are younger than the features that are cut. The third major principle of relative dating is commonly called the **principle of crosscutting relationships.** Geologically speaking, faults or dikes that crosscut or that break series of strata are younger than the deposition of the strata. In some instances, additional horizontal sedimentary beds have been deposited over old fault surfaces, burying them, and providing evidence of the time of origin of the faults.

An example of a fault is shown in figure 6.1A. The black bed, layer 2, in both blocks A and B, was deposited as part of the same originally horizontal sheet. As a result

of faulting, one block has moved down relative to the other, at some time after deposition of the black horizontal bed and the overlying bed 3 rocks. Movement along the fault, the break in the Earth's crust, postdates, or is younger than, deposition of the horizontal crosscut beds. How much younger is impossible to tell from figure 6.1A, because there are no key horizons or beds above the fault that were not broken and that could establish a positive youngest date possible. The fault could have happened at any time after deposition of bed 3, the youngest bed cut by the fault.

Essentially the same relationships are shown in faulted beds in figure 6.1B; but here, after displacement, the fault has been buried by younger horizontal rocks. The overlying layer has not been cut by the fault and hence is younger than the movement of the fault. Relationships shown in figure 6.1B establish that the fault that cuts block A and B and that displaced the black horizontal bed followed deposition of beds 1 to 3, but preceded accumulation of the sediments of bed 4 that bury the fault. Movement along the fault is dated by crosscutting and superposition relationships. Faults and folds can be dated in this general manner, by determination of the youngest rocks involved in the faulting and folding and of the oldest rocks that have not been involved in the movements.

Major unconformities, erosional breaks, or loss of record, can be dated in the same manner. For example, beds 1 to 7 in figure 6.2A were deposited and then tilted or folded as shown in figure 6.2B. The tilted edges were eroded to produce an erosional plane as shown in figure 6.2C; the eroded surface was then buried with rock units 8 and 9 as shown in figure 6.2D. Unit 8 rests on various rock units below the erosional surface or unconformity. On the left side of the block, unit 8 rests on beds 1 and 2, whereas on the right it rests on beds 6 and 7. Many beds have been removed on the left part of the block.

Erosion and formation of the unconformity must be prebed 8, of necessity, because bed 8 has buried the erosional surface, and has left debris of all formations from 1 through 7 strewn across the surface like crumbs on a cake

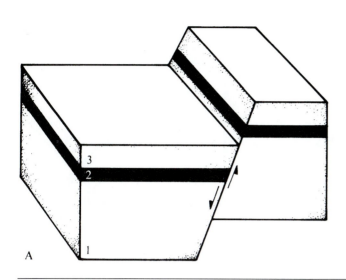

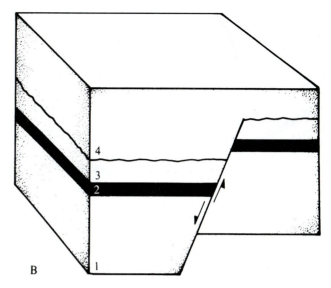

Figure 6.1 Block diagrams showing relationships of normal faulting. **A.** Relationships subsequent to faulting, but prior to deposition. **B.** Relationships after faulting and deposition, which allow relative dating of the fault. Unconformities are shown by irregular wavy lines.

platter. The youngest bed that can be seen below the unconformity in the diagram is bed 7; by the law of superposition, 7 is the youngest bed in the sequence of 1 through 7 in figure 6.2A. Therefore the period of erosion must have followed deposition of bed 7 but preceded deposition of bed 8. The erosional surface then would be dated as post-7 and pre-8.

Two other methods of relative dating can be illustrated in relationship to the character and thickness of the beds involved. In figure 6.3A, the thickness of units 1 through 3 remain relatively constant from the left to the right side of the block diagram. However, unit 4 thins over the small fold in the central part of the diagram and thickens very markedly along the flanks of the fold. This relationship suggests that the folding began sometime during deposition of bed 4, because bed 3 and older units below are folded and the upper surface of bed 4 is relatively horizontal. Differences in thickness between the top and bottom of bed 4 indicate the amount of folding. Bed 5 is parallel to the upper surface of bed 4 and indicates that the minor folding was accomplished before deposition of unit 5. From thickness alone, in fortunate circumstances, one can obtain some information on periods of deformation.

These examples, coupled with unconformities such as in figure 6.3B, are the evidence for dating of the major periods of mountain building that have affected the continental borders of North America during the geologic past.

The nature of the sediments related to erosional surfaces and to fault scarps, or other features of relief, may also aid in defining the relative time of formation of particular features. For example, the clastic wedges of the De-

vonian Catskill delta or of the major Cretaceous belts of coarse conglomerate and sandstone in western North America effectively date the time of major uplift of the Acadian landmass in the east and the Sevier Mountains in the west.

Dating of intrusive igneous rocks is generally based on the principle of crosscutting relationships. One can formulate a general rule and say that intrusive igneous masses are younger than the rocks they invade and older than the rocks that overlie them in a nonconformable or erosional relationship.

A lava flow and a horizontal sill or a sheet of intruded igneous material appear similar on maps but have quite different age relationships. Figure 6.4C shows the relationships seen with a lava flow. The rocks below are baked, shown by the stippled pattern, but the rocks above were deposited over the cooled lava surface. The upper surface may be irregular or it may be eroded to a nearly smooth plane, but debris of the lava commonly occurs in the basal part of the overlying sequence. Lava flows are dated like sedimentary layered rocks, using superposition. Relationships of a sill are shown in figure 6.4D. The sill is shown as the irregularly margined mass with stippled baked zones in the enclosing subparallel beds. Country rock near the igneous mass shows zones of baking because of intrusion of the fluid melted rock material into the country rock. In contrast to lava flows, the country rock both above and below the sill has been baked. Minor fingers or apophyses of the igneous molten material have penetrated into the overlying and underlying rocks.

Dating of some igneous intrusive masses may pose other problems as well in instances where the melt has not penetrated to the surface. Crosscutting relationships date

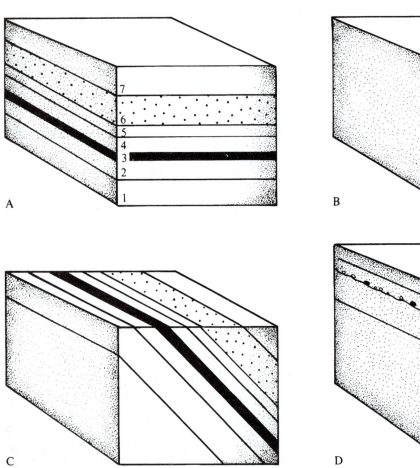
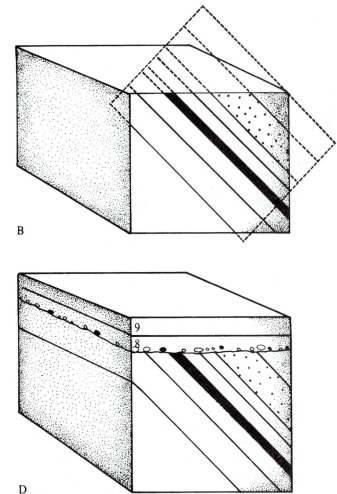

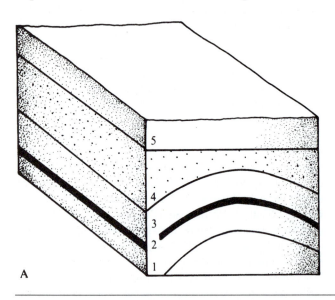
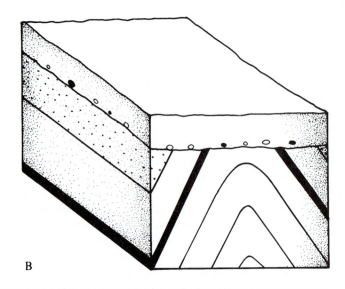

Figure 6.2 Block diagrams showing sequence of development of an angular unconformity. **A.** Deposition of beds 1 to 7. **B.** Tilting or folding of the sedimentary sequence. **C.** Erosion of the tilted beds to produce a horizontal surface. **D.** Burial of the eroded surface by deposition of beds 8 and 9 to produce the angular unconformity.

Figure 6.3 Block diagrams of two folded sedimentary sequences showing different methods of dating. **A.** Folding with contemporaneous deposition demonstrated by variations in thickness of bed 4 without evidence of erosion. **B.** Folded and eroded anticline that was subsequently buried by later deposition. Folding predated deposition of the overlying horizontal bed.

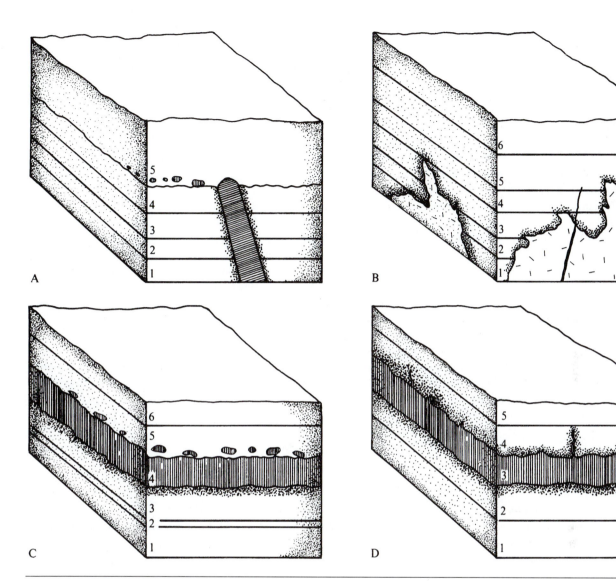

Figure 6.4 Block diagrams showing various relationships of igneous and sedimentary rocks that are useful in relative dating. **A.** A dike has intruded beds 1 through 4 but is overlain unconformably by a younger deposit, bed 5. Eroded remnants of the dike are included in the basal deposits of bed 5. **B.** Two intrusive igneous masses that cut and bake beds 1 through 5. The age relationships of bed 6 and the intrusive masses cannot be determined. **C.** A lava flow, bed 4, has baked part of bed 3 and has supplied erosional debris that is included in bed 5. The flow postdates bed 3 and predates bed 5. **D.** An igneous sill that has baked both the overlying and underlying beds is younger than any bed of the sequence.

the mass only as younger than the youngest rocks affected by the heat or by the intrusive mass. This type of relationship is shown in figure 6.4B where two intrusive masses cut through rock sequences 1 through 5 but have not affected the rocks above bed 5. Both intrusive masses can be dated only as postbed 5. The small black dike postdates the larger intrusive mass that it cuts. A younger limit cannot be definitely established for either.

Compare this with figure 6.4A where the igneous mass was intruded, then exposed during a period of erosion, and then buried by sediments deposited above the unconfor-

mity. In this instance, beds 1 through 4 have been affected by the intrusive mass, but unit 5 above has not been affected. Eroded fragments of the intrusive dike have been incorporated into the base of unit 5 as well. Dating of the intrusive mass can be established in a manner similar to that of the crosscutting relationships seen earlier in figure 6.1. Formation of the intrusive mass in this instance is postbed 4 and prebed 5. The time span between beds 4 and 5 may be great and without other methods, such as will be studied in the exercise on radiometric methods, precision of relative dating is sometimes limited.

Procedure

Part A

1. Using these dating techniques, determine the sequence of geologic events for each of the blocks in figure 6.5. Write the histories in list form, with the oldest at the bottom numbered 1, and youngest at the top. In general, the blocks are arranged from those with the simplest histories to those with the most complex histories. Begin with block A and work toward the more complex blocks. Relationships that are not immediately evident can be most easily interpreted by working from the surface down, or by working backward through time, starting with the youngest event and proceed down into the geologic history recorded earlier in the blocks. Keep in mind the principles of original horizontality, cross-cutting relationships, and superposition, as well as the evidences afforded by sedimentary character and by the degree of folding or metamorphism. Unconformities are shown by irregular wavy lines; faults are shown by relatively thick lines. Erosional debris of older material is shown by symbols suggesting boulders or cobbles of older material in overlying sedimentary sequences. Intrusive contacts are generally shown with a toothed or sharply serrated margin where appropriate. The upper surfaces of several blocks show details or relationships that may not be evident in cross sections on the side of the block. In this same sense, a geologic map may show relationships that have relative time significance, which may not be visible in one of the cross sections along the margin.

Part B

One of the best-kept secrets of the U.S. Park Service is Grand Canyon National Monument. Located approximately 170 kilometers downstream from Grand Canyon National Park, the monument area displays a dramatic close-up view of the Grand Canyon, several episodes of faulting, and a spectacular display of recent as well as ancient volcanic activity including a lava cascade that spilled into the Grand Canyon adjacent to Vulcan's Throne and is preserved now as a frozen lava fall.

At this location, it is possible to apply the rules of relative dating to reconstruct the local geologic history with great clarity. Superposition and crosscutting relations are clearly illustrated by the geologic relationships that can be observed in the rocks of the monument area.

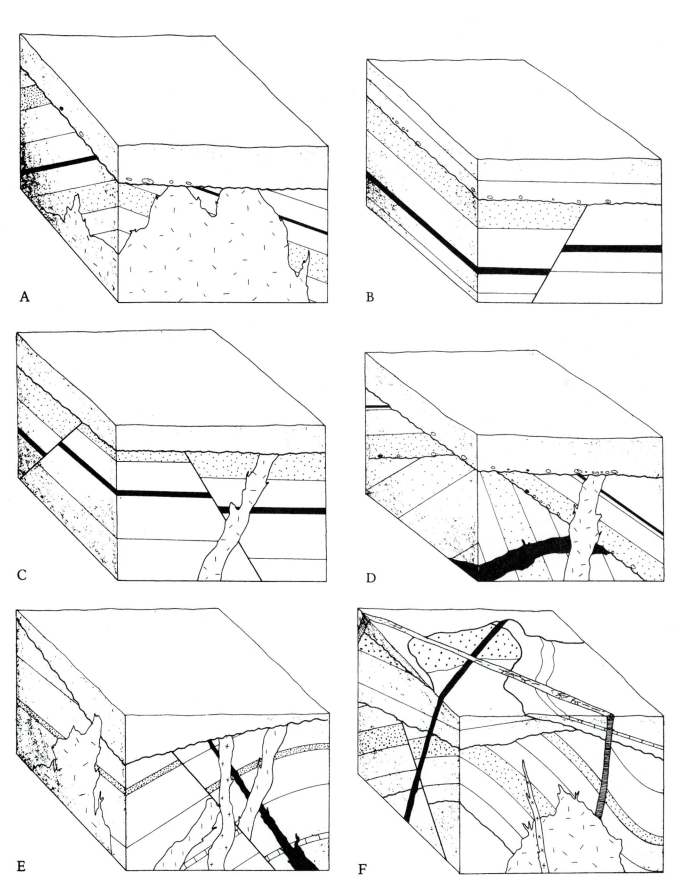

Figure 6.5 Block diagram problems concerning relative dating.

Procedure

Figures 6.6A to 6.6C illustrate, in photo, sketch, and cross section, the rocks in the immediate vicinity of Vulcan's Throne at Grand Canyon National Monument. Study the photograph and the two line drawings of the area and determine the relationships that bear on relative dates for these rocks.

1. Using superposition and crosscutting relations, establish the proper chronologic sequence for the following list of topographic features seen in this area:

Erosion of Grand Canyon

Erosion of ancient Toroweap Valley

Basalt cascades

Deposition of Supai Formation

Deposition of Muav Limestone

Deposition of Redwall Limestone

Deposition of Temple Butte Formation

Most recent eruption of Vulcan's Throne

Early displacement of Toroweap Fault

Late displacement of Toroweap Fault

Deposition of Bright Angel Shale

Basalt filling ancient Toroweap Valley

Lava Dams (remnants of which are preserved against the lower walls of canyon, labeled *Intracanyon Flows* on the cross section)

2. The elevation of the Colorado River below Vulcan's Throne is 1675 feet. How deep was the lake behind the highest lava dam adjacent to Vulcan's Throne? Assume the lava dam was level with its remnant on the canyon wall below Vulcan's Throne.

3. What is the total displacement of the Toroweap Fault in this area?

4. What is the thickness of the lava fill in ancient Toroweap Canyon?

5. If the basalt at the top of ancient Toroweap Valley yields a radiometric date (K/Ar) of 15,000 years, and the lowermost flows in the same valley yield an age of 1.2 million years, calculate the rate of basalt filling in the valley (feet / m.y.).

A

Figure 6.6 **A.** Photograph of Vulcan's Throne and immediate area looking northeast. The difference in elevation from the Colorado River to the top of Vulcan's Throne is 3422 feet. **B.** Sketch of the area included in the photograph. **C.** Line drawing of the wall of Grand Canyon below Vulcan's Throne. The basalt cascade shown in these illustrations is derived from a volcano located between 4 and 5 miles north-northwest of Vulcan's Throne. This view is looking to the north. (Photograph and drawings courtesy of W.K. Hamblin.)

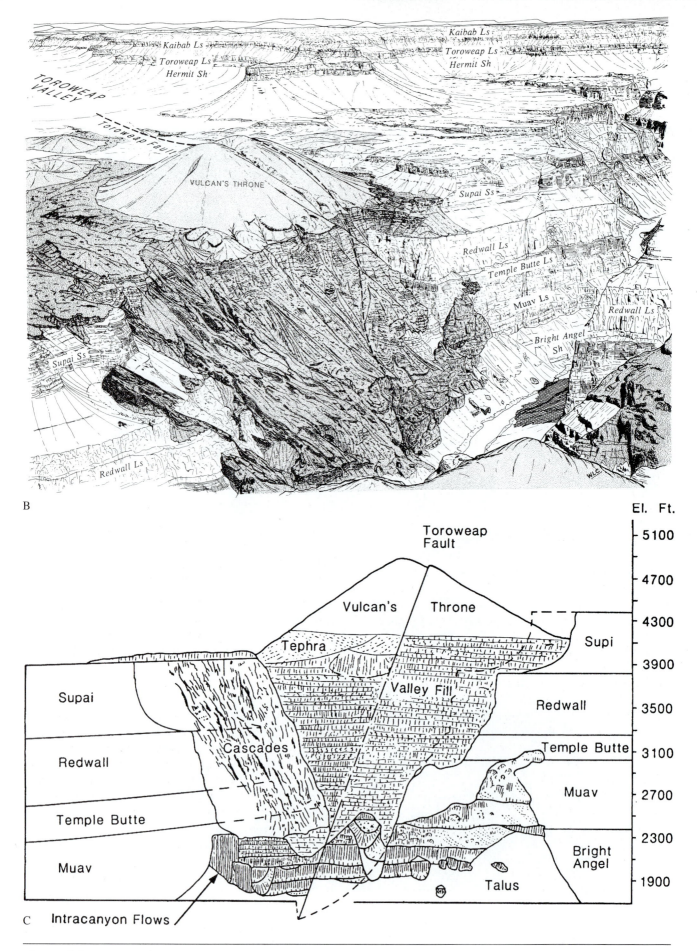

B

C   Intracanyon Flows

Figure 6.6 continued

Radiometric Dating

In 1896, a French physicist, Henry Becquerel, discovered radioactivity, a process whereby atoms of certain elements, uranium for example, spontaneously break down to form atoms of new and lighter elements such as lead. Using the principle of radioactivity, an American chemist, Bertram Boltwood, in 1907, studying the decay of uranium to lead, calculated ages for rocks in various parts of the geologic column. In spite of his rough calculations because of lack of accurate knowledge of the decay rate of uranium, Boltwood's computed dates were remarkably close to presently accepted dates for the same rocks. The presently accepted age of the Earth is approximately 4.6 billion years. This date is drastically older than Earth ages calculated by other methods used prior to the discovery of radioactivity.

It is believed that the salt in the sea has been derived from weathering of rocks on the Earth's surface. In 1899, John Joly attempted to arrive at an approximate age of the Earth by measuring the amount of salt being carried annually to the sea by rivers of the world and by comparing that to the total salt in the ocean. Joly estimated 90 million years had elapsed since the first freshwater oceans condensed on the Earth's surface.

Several scientists have approached the age of the Earth by studying the thickness of sedimentary rocks on the Earth's surface and by comparing them with modern rates of accumulation for these rock types. The computed ages of the Earth, using rates of deposition, range from 17 million years to as much as 1.5 billion years. A major weakness of this approach, like the weakness of the approach by Joly, is that the rates of accumulation of sedimentary rocks, or salt transfer, is poorly known and highly variable from year to year; thus, conclusions based upon these methods are inaccurate.

Lord Kelvin approached this problem by studying cooling of the Earth. He assumed that, in the beginning, the Earth was a molten mass and, by calculations involving rate of heat loss at the Earth's surface, concluded that the Earth was approximately 24 million years old. Kelvin neglected to evaluate heat generated within the Earth since its formation, such as the heat generated by radioactive decay, and hence his method is of limited value.

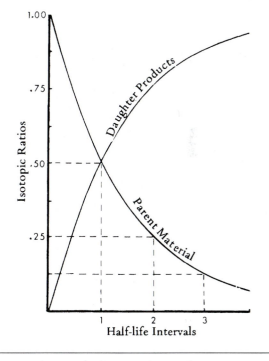

Figure 7.1 Graph showing decay of parent material with simultaneous buildup of daughter products over half-life intervals of time.

Radiometric dates are calculated in number of years and are referred to as absolute dates in contrast to geologic dates based upon reference standards, or relative dates. The basis for radiometric dating is radioactivity, the spontaneous decay of certain isotopes, called **parent isotopes,** to produce unique end products called **daughter products,** which accumulate as the parent material disintegrates as shown in figure 7.1.

Most radioactive isotopes decay by one of the following types of radiation:

Alpha radiation: An alpha particle is a nuclear particle composed of two protons and two neutrons. The loss of an alpha particle, therefore, reduces the parent material by four mass units (atomic weight) and reduces the atomic number by two. That is, an element X whose weight is w, and number is n ($_n^wX$) would change, following the emission of an alpha particle, to become another element Y whose weight and number would be $_{n-2}^{w-4}$. An example of alpha decay is the initial step in the radioactive series U-238–Pb-206, in which the parent, uranium-238, is changed to thorium-234. The complete transition to lead-206 involves 8 alpha and 6 beta emissions.

Beta radiation: Beta decay is a high-energy electron released from the nucleus converting a neutron into a proton. The parent material changes only by the addition of a single proton in its nucleus, or an atomic number change of +1 ($_n^wX \overset{beta}{\rightarrow} _{n+1}^w Y$). Mass number remains the same because the mass of the emitted electron is insignificant compared to the masses of the neutrons and protons. Beta radiation can also result by capture of an electron by the nucleus. The result of this change is the same except the atomic number decreases by 1 ($_n^wX \overset{beta}{\rightarrow} _{n-1}^w Y$). Rubidium-87 in decaying to strontium-87 is an example of losing an electron and increasing the atomic number by 1. Potassium-40 decaying to argon-40 illustrates electron capture and a decrease in the atomic number by 1.

Gamma radiation: Energy release is similar to X rays. The decay rate, as well as the type of radioactive particle emitted, is unique for each radioactive isotope. The time period required for half the atoms to decay is called the **half-life.**

Radiometric methods of age determination have limitations. Measured ages are commonly within 5% of the true age. Some measurements are more precise, some less, depending upon a variety of conditions. Some of the problems introduced into the calculations of radiometric dating include:

1. Samples used in radiometric dating must have remained a closed system from the time of their crystallization until the time of measurement. Significant amounts of material must not have been added by subsequent deposition or removed by solution or other mobilization, otherwise the calculated ages will be erroneous and misleading. When selecting a sample, it is of the utmost importance to choose rocks that have every indication of having been unaltered since their initial formation. The closed system requirement is in most instances the weakest part of the age determination.

2. Measuring processes in the laboratory require extreme care and accuracy. Errors made in the laboratory, with a few exceptions such as measurement of ages of young rocks by potassium/argon, are generally small.

3. No initial daughter is present or, if present, must be corrected for. The corrections are routinely done in Rb/Sr, U/Pb, and usually unnecessary in K/Ar dating.

4. The geologic control criterion infers that any date must relate to a field problem, and on completion of the analyses, must help in the solution of said problem. Otherwise a meaningless (and expensive) number, not a

date, is produced. Good geologic interpretation of one or more dates is required to produce an *age.*

Figure 7.2 illustrates the radioactive isotopes, their stable end products, and the half-lives of the parent isotopes most commonly used for radiometric dating. As shown in figure 7.3, uranium-238 ultimately decays to lead-206. Similarly, potassium-40 decays to argon-40 (fig. 7.4), and rubidium-87 to strontium-87. The amount of daughter product compared to the amount of parent material is a function of duration of time during which this transmutation process has been operative. In other words, as time continues there is an accumulation of daughter material at the expense of the parent, and a measurement of the ratio of daughter to parent material will allow the calculation of the age of the sample. If the sample has been altered, or if solutions have removed any material, the system becomes open and the calculated age is invalid.

The principle of radiometric age dating can be expressed by the following simplified relationship:

$$t = \frac{Nd}{Np\lambda}$$

where t = time in years, Nd = amount of daughter products, Np = parent material, and λ = rate of decay per year, or decay constant $\lambda = \dfrac{0.693}{\text{half-life}}$. The amounts of daughter and parent material can be measured, λ is known, and the equation can be solved for t.

One expression of the general formula for solution of radiometric dating is:

$$t = \frac{1}{\lambda} \log e\left[\frac{D_p - D_i}{P_p} + 1\right]$$

where D_p = amount of daughter product at present time
D_i = amount of daughter product existing initially
P_p = amount of parent material at present time
t = time in years since crystallization of sample material

Because potassium-40 decays into two daughter products, calcium-40 and argon 40^{rad}, the general formula for calculating time must be modified. The potassium-argon age equation is:

$$t = \frac{1}{\lambda_e + \lambda_b} \log e\left[\frac{^{40}_{rad}Ar}{^{40}K}\left(\frac{\lambda_e + \lambda_b}{\lambda_e}\right) + 1\right]$$

where λ_e = decay constant for radiogenic argon-40
λ_b = decay constant for calcium-40

Potassium-argon methods are particularly useful because they are effective in dating the oldest rocks on Earth (3.8 billion years) and those as young as 100,000 years.

Isotopes	Half-life	λ	Effective Dating Range	Source Materials
$^{238}U/^{206}Pb$	4.46×10^9 years	1.55×10^{-10}	>100 m.y.	Zircon, Sphene, Apatite whole rock
$^{40}K/^{40}Ar$	1.25×10^9 years	$\lambda\epsilon = 0.581 \times 10^{-10}$ $\lambda\beta = 4.96 \times 10^{-10}$	>100,000 years	Biotite, Muscovite, Hornblende, Glauconite, K-feldspar, Bentonite whole volcanic rock
$^{87}Rb/^{87}Sr$	4.88×10^{10} years	1.42×10^{-11}	>100 m.y.	Biotite, Muscovite, K-feldspar some clay minerals, whole rock (granitic), Apatite, clay-rich shales
C-14	5730 years		<50,000 years	Organic materials

Figure 7.2 Chart showing various isotopes used in radiometric dating, their half-lives, decay constants, effective dating ranges, and source materials.

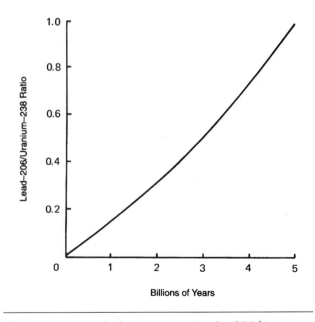

Figure 7.3 Graph showing variation lead-206/ uranium-238 ratio with respect to time.

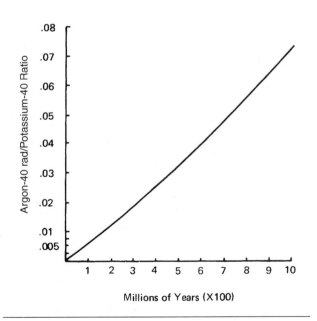

Figure 7.4 Graph showing the variation of argon-40 rad and potassium-40 with time in a closed system.

This method has further application in dating marine sedimentary rocks containing glauconite, a secondary potassium-bearing mineral, as well as the commonly used minerals and whole volcanic rocks.

Because of the extremely long half-life of rubidium-87, the practical formula for the calculation of ages using the strontium/rubidium ratio is:

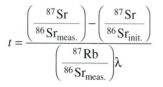

$$t = \frac{\left(\dfrac{^{87}Sr}{^{86}Sr_{meas.}}\right) - \left(\dfrac{^{87}Sr}{^{86}Sr_{init.}}\right)}{\left(\dfrac{^{87}Rb}{^{86}Sr_{meas.}}\right)\lambda}$$

On the same principle, Willard Libby, the Nobel Prize–winning chemist, described the carbon-14, or radiocarbon dating methods, useful for measuring the time interval 500–50,000 years. Nitrogen-14 is bombarded by cosmic rays in the upper atmosphere and converted to carbon-14, a unique isotope of carbon. Carbon-14 is radioactive with a half-life of 5730 ± 40 years. The rate of production of carbon-14 in the atmosphere is assumed to be balanced by the rate of the decay as radioactive carbon-14 changes back to its parent, nitrogen-14. Therefore, at any one time, the atmosphere is in equilibrium with a constant amount of carbon-14, which is oxidized to carbon dioxide.

The newly formed carbon-14 mixes into the atmosphere and oceans where it is utilized in the production of organic compounds and becomes a part of the food chain. Living organisms continually replenish their C-14 supply by food intake; therefore, only after death is there a depletion of C-14 because of radioactive loss. The carbon-14 clock for any organism, therefore, "starts," or actually begins to disintegrate, upon death. The longer the time lapse since death, the lower will be the amount of C-14 within the remains. Carbon-14 radiation can be measured by a special type of Geigertube, and with extreme care in sample selection, preparation, and shielding against background radiation, measurements can be obtained on samples up to 50,000 years in age, although accuracy rapidly diminishes beyond 30,000 years.

As an evaluation of the stability of the environment over long intervals of time with regard to C-14, Libby and his associates, Arnold and Anderson, sampled organic materials from archaeological sites of reasonably definite ages ranging back 5000 years. Comparisons of historical and radiocarbon-calculated dates are shown in figures 7.5 and 7.6. The high degree of correlation between these two approaches is a verification that the ratio of C-14 in the atmosphere has remained constant over the past 5000 years. On the basis of this evidence, it is assumed that the atmosphere composition has remained constant, with respect to C-14, over the entire possible dating interval of approximately 50,000 years.

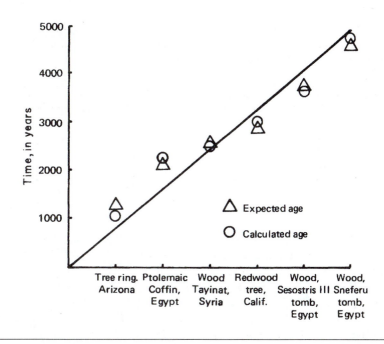

Figure 7.5 Graph showing comparison of expected ages and calculated ages of ancient pieces of wood, using carbon-14 content. (Data from Libby.)

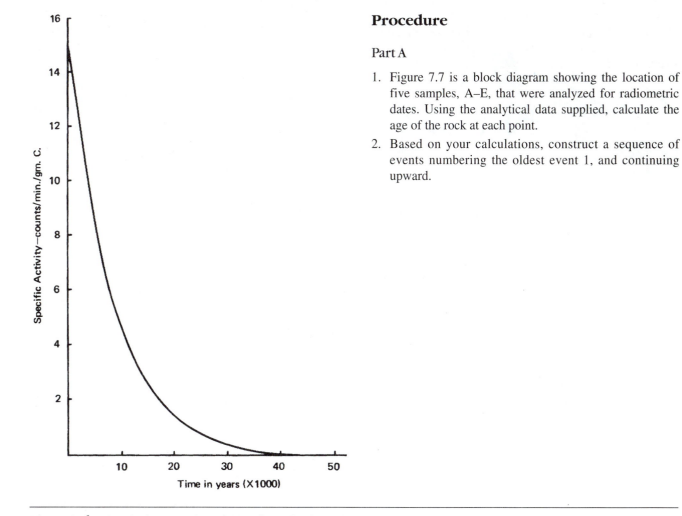

Procedure

Part A

1. Figure 7.7 is a block diagram showing the location of five samples, A–E, that were analyzed for radiometric dates. Using the analytical data supplied, calculate the age of the rock at each point.

2. Based on your calculations, construct a sequence of events numbering the oldest event 1, and continuing upward.

Figure 7.6 Graph showing the relationship of radioactive carbon-14 and time, based upon a half-life of 5730 years.

Sample	Analysis	Age	Geologic Column
A (Zircon)	Of the total of these two isotopes, 22% is Pb-206, and 78% is U-238		
B (Biotite)	$^{40}_{rad}Ar/^{40}K$ ratio 0.030		
C (Biotite)	$^{87}Sr/^{86}Sr$ meas. 0.75 $^{87}Sr/^{86}Sr$ init. 0.70 $^{87}Rb/^{86}Sr$ meas. 5.0		
D (Wood)	Specific activity 1 ct./m/g		
E (Bone)	Specific activity 8 cts./m/g		

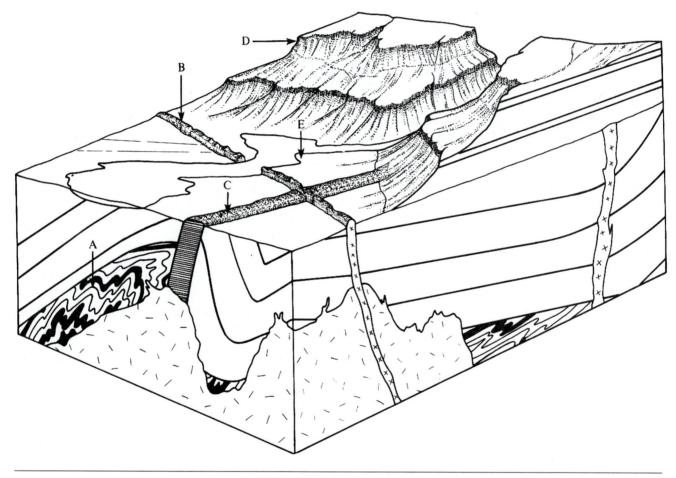

Figure 7.7 Block diagram problem showing the collecting localities of samples **A** through **E.**

EXERCISE 8

Continental Drift and Plate Tectonics

From early concepts of continental drift by Taylor (1910) and Wegener (1912), the modern theory of plate tectonics, or global tectonics, has evolved. Geologists now explain the mechanism of geologic change in relation to at least 10 major plates making up the Earth's surface. These plates are approximately 100 km thick and move over the Earth's surface at an average velocity of 2–4 cm per year.

The major development of the Earth's surface features has been, and is now, taking place along plate margins where adjacent plates interact. Plate margins can be classified according to relative motion, which produces three types: divergent, plates moving apart; convergent, plates moving together; and transform faults, horizontal side-by-side movement. A further subdivision can be described based upon the type of major physiographic feature at the plate margin such as continent–continent, where two continental masses are opposed along opposite sides of plate boundaries; continent–ocean basin, a continent adjacent to an ocean basin; and ocean basin–ocean basin, a plate margin within a single oceanic basin. The combination of plate motion and physiographic character is a reasonable means of analyzing Earth dynamics.

Procedure

Part A

Outline maps of South America, Africa, Europe, and North America are given in figure 8.1. The stippled area around each of the continents represents the area between the present shoreline and the 500-fathom (900-m) line. The latter is approximately the edge of the granitic continental crustal margin.

A fitting of the continents is best done on a small globe but various map projections have been used by others. Because of the limitations of representing a curved surface on a plane, most map projections distort either the shape or size of the continents and alter their margins. Because of these difficulties, a modified mercator projection has been used in the present exercise. This projection minimizes distortion of the coastline to be fitted. Various sections of Central America and Europe have been shifted slightly out of their present-day positions, but these are areas of relatively recent orogeny and may have had different positions in the past.

1. Trace the continental outlines following their true margin at the outer edge of the continental shelf from figure 8.1. Adjust their relative positions until the best fit is achieved, not only of continental outlines but also of geologic data, which are summarized on some of the maps.

Part B

Available evidence suggests that until the Late Paleozoic, the continents of Africa and South America were essentially side by side, and that drifting and opening of the Atlantic Ocean began early in the Mesozoic. If the continents drifted and the Atlantic Ocean is still enlarging, there must be some evidence on the present ocean floor. Various workers studied the topography, sediments, and igneous rocks of the mid-Atlantic ridge to see if there are clues or evidence to document movement. It soon became apparent that a major linear topographic depression, or rift zone, is superimposed along the crest of the mid-Atlantic ridge along most of its length. Not only was the rift recognized in the Atlantic, but the linear topographic low appears to be part of a major series of rifts and ridges associated with mid-oceanic areas of basaltic volcanic activity and seismic activity. The distribution of this worldwide system is shown in figure 8.2.

The topographic and structural features were documented at about the same time that parallel belts of alternating reversed and normal polarity of the magnetic intensity, or magnetic field strength, were recognized in rocks along the ridge trend. These alternating bands of normal and reverse polarity in the basalts occur symmetrically along the

73

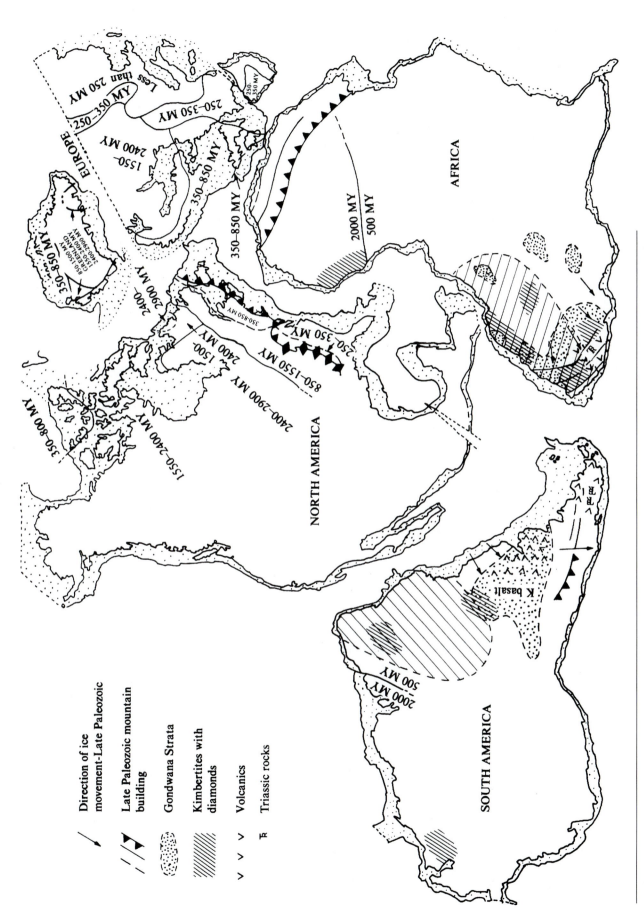

Figure 8.1 Maps of outlines of South America, North America, Africa, and Europe showing positions and ages of various major mountain belts. Ruled pattern shows approximate range of late Pennsylvanian-early Permian aquatic reptile *Mesosaurus* in South America and Africa.

77

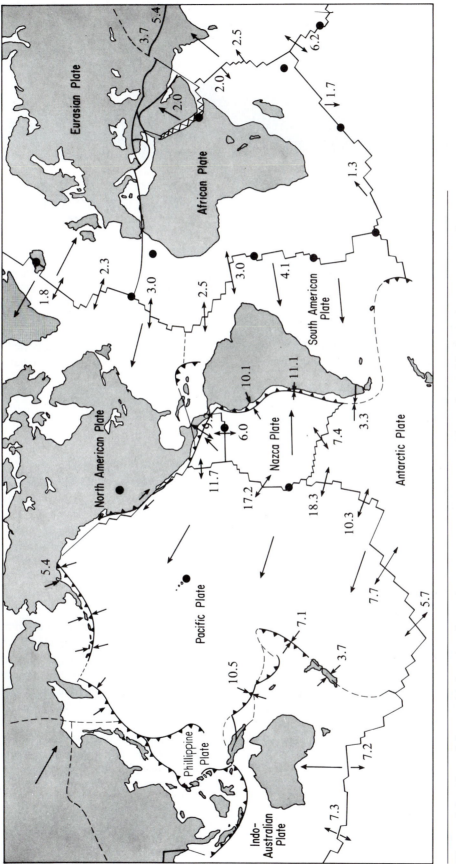

Figure 8.2 Map of the world showing the major plates, their boundaries, and direction of motion. Subduction zones are indicated by barbed lines, barbs are on overriding plates. Black circles are presently active hot spots. Numbers indicate rate of motion in cm/yr. (Rates after McKenzie and Richter, 1976.)

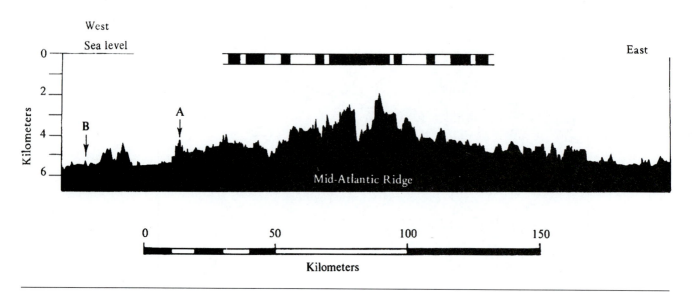

Figure 8.3 A schematic cross section of the central part of the mid-Atlantic ridge showing topography and patterns of paleomagnetic data. The dark and light areas correspond to the pattern shown on figure 4.6.

rift zone in the Atlantic Ocean in the fashion illustrated in the generalized diagram in figure 8.3. The dating of sequences of polarity reversals is shown in figure 4.6.

1. Why are the belts of polarity contrast situated symmetrically on either side of the rift zone on the mid-Atlantic ridge?

2. Calculate as closely as possible the rate, in centimeters per year, at which sea floor spreading is occurring in the Atlantic Ocean, based upon the paleomagnetic data. Your calculation is the rate at which a single plate is moving.

3. Estimate the time necessary for the spread of the south Atlantic Ocean between Africa and South America by dividing the present separation by the rate of spreading as calculated in question 2. Note: This answer requires doubling the answer from question 2.

4. What ages of basalt might be expected at points A and B, assuming the rate of spreading to have been constant?

Part C

The chart on page 79 shows the three major plate margin types as well as the three possible physiographic combinations for each. By studying figure 8.2, complete the chart by entering names of modern geographic locations that serve as examples for each of the nine possible combinations shown on the chart.

Part D

Seismic or earthquake data provide us with direct information concerning plate interactions or tectonics. The U.S. Geological Survey in Washington, D.C., maintains and publishes a monthly record of seismic events around the world. These records, distributed to the public by the National Earthquake Information Service, are entitled Preliminary Determination of Epicenters. They include date; time; location of epicenter; depth of focus, or hypocenter; and magnitude of every earthquake of magnitude 3 or greater occurring in the United States, and of those greater than

Modern Examples by Geographic Name

Plate Margin	Characteristics	Continent–Continent	Continent–Ocean Basin	Ocean Basin–Ocean Basin
Divergent	Tension Crustal lengthening (plate generation) Normal faults Shallow earthquakes Basaltic vulcanism			
Convergent	Compression Crustal Shortening (plate destruction) Reverse and thrust Faults Folds Shallow and deep earthquakes			
Transform Fault	Lateral movement Strike-slip or horizontal moving faults Shallow earthquakes			

Table 8–1 The Modified Mercalli Earthquake Intensity Scale

Scale Degree	Effects on Persons	Effects on Structures	Other Effects	Rossi–Forel Equivalent	Equivalent Shallow Magnitude
I.	Not felt except by few under favorable circumstances.			I.	
II.	Felt by few at rest.		Delicately suspended objects swing.	I–II.	2.5
III.	Felt noticeably indoors. Standing cars may rock.		Duration estimated.	III.	
IV.	Felt generally indoors. People awakened.		Cars rocked. Windows, etc., rattled.	IV–V.	3.5
V.	Felt generally.	Some plaster falls.	Dishes, windows broken. Pendulum clocks stop.	V–VI.	
VI.	Felt by all. Many frightened.	Chimneys, plaster damaged.	Furniture moved. Objects upset.	VI–VII.	
VII.	Everyone runs outdoors. Felt in moving cars.	Moderate damage.		VIII.	5.5
VIII.	General alarm.	Very destructive and general damage to weak structures. Little damage to well-built structures.	Monuments, walls down. Furniture overturned. Sand and mud ejected. Changes in well water levels.	VIII–IX.	6
IX.	Panic.	Total destruction of weak structures. Considerable damage to well-built structures.	Foundations damaged. Underground pipes broken. Ground fissured and cracked.	IX.	
X.	Panic.	Masonry and frame structures commonly destroyed. Only best buildings survive. Foundations ruined.	Ground badly cracked. Rails bent. Water slopped over banks.	X.	
XI.	Panic.	Few buildings survive.	Broad fissures. Fault scarps. Underground pipes out of service.	X.	8.0
XII.	Panic.	Total destruction.	Acceleration exceeds gravity. Waves seen in ground. Lines of sight and level distorted. Objects thrown in air.	X.	8.5

Source: From Smith, P. J., 1973. *Topics in Geophysics*, M.I.T. Press, p. 176.

magnitude 4 in other places in the world. Further information is included for some earthquakes. Occasional comments include the intensity as measured on the Modified Mercalli earthquake intensity scale, which is illustrated in table 8–1. This kind of information is extremely useful for understanding the dynamic nature of our planet. By studying this information, a student can gain firsthand knowledge of the frequency, distribution, and energy of the Earth's seismic activity.

The epicenters of the earthquakes occurring during the entire month of January 1982 have been plotted on a world map and are illustrated as figure 8.4. In addition, the seismic activity for the week 24–31 December 1981, is presented in table 8–2. Abbreviations used in the headings are:

UTC = universal time; GS = geological survey data; MB = magnitude based on "P" seismic waves; and MSZ = magnitude based on vertical surface waves. An asterisk (*) following the time indicates a less reliable solution in the calculation because of incomplete or less reliable data.

1. Using a colored pen or pencils, plot every other epicenter from table 8–2 on figure 8.4. When you are finished, your map will display essentially five weeks of seismic activity.

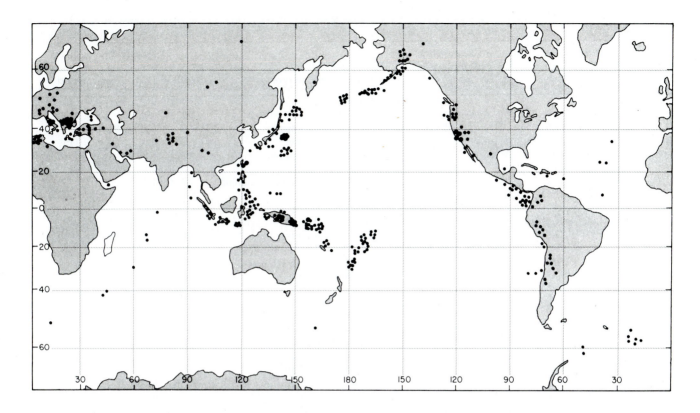

Figure 8.4 Map of the world showing the location of epicenters of seismic events for the month of January 1982. Small dots represent single earthquakes, large dots represent 5 seismic events that occurred in a single location, at least within the scale of this map.

2. Compare your finished map (fig. 8.4) with figure 8.2. Is there a clear correlation between the distribution of recent earthquakes and the boundaries of plates? Is such a correlation expected?

3. Seismic foci, or hypocenters, are classified as follows: shallow occur to depths of 50 km, intermediate to 250 km, and deep in excess of 250 km. What is the pattern in the distribution of intermediate and deep foci earthquakes reported in table 8–2? What controls this pattern?

4. The earthquake of 31 December 1981, near the Bonin Islands region, is a common site for deep earthquakes. Why is this?

5. What is the probable cause of the Kazakh seismic event of 27 January at 03:43 hours? What two things form the basis of your answers?

Table 8–2 Preliminary Determination of Epicenters

Day, UTC 12/81	Hr.	Min.	Sec.	Lat.	N/S	Long.	W/E	Region and Comments	Depth	MB	MSZ	No. Sta. Used
24	02	22	07.7	34.017	N	116.767	W	Southern California	20			10
24	04	24	53.6*	14.641	N	119.919	E	Luzon, Philippine Islands	33	4.6		8
24	05	33	21.5	29.956	S	177.701	W	Kermadec Islands	33	6.1	6.8	151
24	09	43	51.5*	30.041	S	177.563	W	Kermadec Islands	33	5.0	5.1	15
24	11	11	17.1*	22.061	S	175.916	W	Tonga Islands Region	63	4.9		14
24	13	02	40.4	29.925	S	177.374	W	Kermadec Islands	33	5.3	5.4	29
24	14	07	39.3*	39.952	N	77.366	E	Southern Sinkiang Prov., China	33	4.9		7
24	14	44	07.4*	14.282	S	74.321	W	Peru	108			8
24	19	44	53.1*	30.228	S	177.378	W	Kermadec Islands	33	4.9		17
24	22	00	51.6*	33.153	N	49.705	E	Western Iran	33	4.6	4.0	28
24	22	36	28.5	30.076	S	177.387	W	Kermadec Islands	36	5.2	5.4	40
25	00	06	09.5*	12.555	N	88.303	W	Off Coast of Central America	33	4.7		26
25	00	28	16.8	4.738	N	118.458	E	Kalimantan	52	5.4	5.2	47
25	04	55	52.7*	17.406	N	61.768	W	Leeward Islands	49	4.2		12
25	06	03	07.9*	77.013	N	6.601	E	Svalbard Region	10	4.4		14
25	08	28	48.0	6.561	N	73.039	W	Northern Colombia	195	4.5		13
25	09	12	06.4	30.313	S	177.489	W	Kermadec Islands	33	5.4	5.4	57
25	10	37	10.7*	30.200	S	177.411	W	Kermadec Islands	33	4.7		12
25	10	48	45.1*	30.324	S	177.301	W	Kermadec Islands	33	5.2		20
25	12	35	49.6*	11.172	N	62.474	W	Windward Islands	102	4.8		55
25	15	27	18.7*	13.613	S	76.425	W	Near Coast of Peru	33			5
25	15	50	33.3*	23.197	N	121.642	E	Taiwan	33	4.5		11
25	17	02	35.5	53.884	N	160.800	E	Near Coast of Kamchatka	33	4.6		19
25	22	26	41.0*	59.871	N	152.716	W	Southern Alaska	113	4.3		14
26	03	42	19.5*	37.952	N	22.702	E	Southern Greece	33	4.0		25
26	10	17	16.6*	2.209	S	139.847	E	Near N. Coast of West Irian	33	4.8	4.5	18
26	11	16	05.8	23.942	S	66.512	W	Jujuy Province, Argentina	207	4.9		49
26	14	29	11.1*	38.950	N	25.310	E	Aegean Sea	10	4.2	3.6	39
26	17	05	32.8	29.812	S	177.854	W	Kermadec Islands	33	6.3	7.1	149
26	17	53	30.6*	29.983	S	177.830	W	Kermadec Islands	33	5.2		17
26	17	53	38.4*	40.213	N	28.778	E	Turkey Felt in the Istanbul area	27	4.3		28
26	19	38	07.7*	9.773	S	119.218	E	Sumba Island Region	100	4.4		10
26	21	50	43.9	7.260	S	129.183	E	Banda Sea	150	5.3		21
26	22	02	18.5*	22.765	S	68.332	W	Northern Chile	119	4.8		11
27	03	43	14.1	49.923	N	78.876	E	Eastern Kazakh SSR	0	6.1	4.3	167
27	06	22	50.3*	30.063	S	177.622	W	Kermadec Islands	33	5.4	4.9	18
27	10	30	44.4	2.160	S	139.825	E	Near N. Coast of West Irian	33	5.6	5.9	60
27	13	25	33.3*	46.331	N	16.832	E	Yugoslavia	10			13
27	16	36	48.3*	5.090	S	139.311	E	West Irian	33	3.7		6
27	17	39	16.7	39.004	N	24.799	E	Aegean Sea Ten houses damaged on Evvoia. Felt strongly in eastern Greece. Also felt in the Izmir, Turkey area	33	5.3	6.5	112

Table 8–2 continued

Day, UTC 12/81	Hr.	Origin Time, UTC Min.	Sec.	Lat.		Long.		Region and Comments	Depth	Magnitudes, GS MB	MSZ	No. Sta. Used
27	20	21	05.9*	7.107	N	73.093	W	Northern Colombia	138	4.6		10
27	20	24	15.5*	34.250	N	117.617	W	Southern California	9			9
27	21	23	13.6	8.278	S	79.875	W	Near Coast of Northern Peru	33	5.2	4.4	41
28	01	53	02.2	6.889	S	130.004	E	Banda Sea	131	4.6		13
28	10	28	15.9*	54.642	N	160.380	W	Alaska Peninsula	33			9
28	12	40	18.4	14.974	S	168.121	E	Vanuatu Islands	33	5.7	5.2	89
28	13	08	26.2	21.644	N	143.470	E	Mariana Islands Region	33	5.3	5.0	57
28	14	18	15.5*	21.495	N	143.583	E	Mariana Islands Region	33	5.0		30
28	14	38	20.9*	21.893	N	144.109	E	Mariana Islands Region	33	4.5		22
28	14	49	40.6	35.016	N	45.934	E	Iran–Iraq Border Region	33	5.0	4.0	69
28	15	20	11.4*	21.353	N	143.818	E	Mariana Islands Region	33	4.6		13
28	15	40	02.1*	21.642	N	143.718	E	Mariana Islands Region	59	5.1		25
28	16	11	00.4*	14.213	N	92.200	W	Near Coast of Chiapas, Mexico	68	4.7		7
28	16	37	35.8*	21.360	N	143.664	E	Mariana Islands Region	33	4.7		15
28	17	23	52.2	21.511	N	143.789	E	Mariana Islands Region	27	4.9		35
28	17	40	49.8	21.593	N	143.518	E	Mariana Islands Region	33	5.3		36
28	18	10	57.7	13.805	N	95.915	E	Andaman Islands Region	33	5.0	5.1	48
28	18	13	25.5*	21.487	N	143.500	E	Mariana Islands Region	33	5.0		28
28	20	56	01.0*	21.429	N	143.523	E	Mariana Islands Region	33	5.0		29
28	22	01	50.1	63.111	N	150.819	W	Central Alaska	151			13
28	22	45	42.1	37.211	N	114.980	W	Southern Nevada	5			16
								Felt (IV) at Las Vegas. Felt in Clark and Lincoln Counties, Nev. Also felt at Toquerville, Utah, and Temple Bar, Arizona.				
28	23	17	50.8*	21.489	N	143.621	E	Mariana Islands Region	33	4.7		11
28	23	35	59.6*	21.645	N	142.958	E	Mariana Islands Region	33	4.7		13
29	02	22	59.7	21.352	N	143.152	E	Mariana Islands Region	33	4.8		26
29	05	07	22.3*	21.337	N	143.680	E	Mariana Islands Region	33	4.6		14
29	06	39	29.8*	21.474	N	143.694	E	Mariana Islands Region	33	4.7		16
29	08	00	44.9	38.796	N	24.720	E	Aegean Sea	10	4.7	5.3	83
								Felt in the Khalkis–Thessaloniki area				
29	09	26	36.0*	21.425	N	143.841	E	Mariana Islands Region	33	4.8		13
29	15	37	03.9*	21.858	N	143.581	E	Mariana Islands Region	33	4.7		12
29	15	38	26.1*	6.066	S	155.274	E	Solomon Islands	150	5.0		12
29	16	11	17.0*	21.459	N	143.643	E	Mariana Islands Region	33	4.6		9
29	16	37	17.9*	19.209	S	68.168	W	Chile–Bolivia Border Region	205	5.0		9
29	19	06	31.3	30.231	S	177.924	W	Kermadec Islands	60	5.5		36
29	22	49	00.4	21.508	N	143.379	E	Mariana Islands Region	33	5.3	4.8	56
29	23	37	53.3*	21.539	N	143.648	E	Mariana Islands Region	33	4.9		9
30	11	26	36.0*	43.756	N	147.678	E	Kuril Islands	33	5.2	4.1	56
30	13	47	27.3	64.589	N	148.080	W	Central Alaska	33	3.9		11
30	14	00	33.8	64.577	N	148.158	W	Central Alaska	27	4.8		37
								Felt (V) at Ester and (IV) at Fairbanks				
30	15	00	53.8*	22.028	N	143.536	E	Volcano Islands Region	79	4.7		13
30	16	46	34.4*	38.812	N	20.803	E	Greece	33	4.4		11
30	17	44	09.6	6.734	N	126.957	E	Mindanao, Philippine Islands	77	5.0		31
30	20	32	36.1*	13.570	N	90.620	W	Near Coast of Guatemala	33	4.5		12
30	21	09	54.0*	4.349	N	126.008	E	Talaud Islands	104	4.7		12
31	05	08	13.5	27.694	N	139.680	E	Bonin Islands Region	494	4.7		38
31	06	54	51.1*	33.935	S	179.331	W	South of Kermadec Islands	33	5.2		28
31	09	20	05.9*	0.796	N	123.883	E	Minahassa Peninsula	294	5.1		20
31	12	15	54.4	61.910	N	151.758	W	Southern Alaska	128	4.1		14
								Felt at Wasilla and Houston				
31	13	48	43.0*	2.150	N	126.451	E	Molucca Passage	125	4.9		14
31	21	38	13.9	21.448	N	143.716	E	Mariana Islands Region	33	5.0		12

Fossils and Fossilization

Fossils, as remains of once-living organisms, are equal to rocks in terms of their importance as documents of Earth history. The term **fossil** has been variously used, but is here interpreted to be any evidence, direct or indirect, of the existence of organisms in prehistoric time. The term *fossil* comes from the Latin word *fodere,* which means to dig, or *fossilis,* which means dug out or dug up. An exact definition as to what constitutes a fossil, or an upper limit of prehistoric time, is difficult to establish. For example, are the hollow molds of a mule or a man buried in ash in the city of Pompeii fossils, or is the well-tanned and pickled Irish man, buried in a peat bog in the twelfth century, a fossil? Are the remains of a mastodon or a hairy mammoth fossils, when they are still well-enough preserved that the meat can be eaten rather palatably, even though the animals lived in the Pleistocene? The upper fringe of prehistoric time is a difficult and sometimes arbitrary area in the definition of the term *fossil.*

Evidence of fossils may range from exceedingly well preserved forms like the mastodons frozen in the tundra of Alaska and Siberia, to less well preserved forms such as clam shells that have been replaced and recrystallized. Direct evidences of fossils usually tell us much about the shape of the animal or plant and commonly contain materials that were actually precipitated by the life processes of the organism. Indirect fossils on the other hand, are normally only impressions, yet still suggestive of size and proportions. A track, a trail, a burrow-filling, a "stomach stone," or a coprolite are indirect evidences of organic activity and tell us something about the organisms. It has been suggested that perhaps only one organism in a million, living in even a preservable environment, is likely to be fossilized.

Several factors are normally requisite in order that fossilization may take place. First, a body to be fossilized ordinarily must contain **hard parts,** such as teeth, bones, or shells. Hard parts are much more easily preserved, because of their relatively inert nature compared to soft tissues. Soft-bodied forms are occasionally preserved, because the impressions of jellyfish, leaves, tentacles of octopuslike organisms, and trails of gastropod feet have all been noted in the geologic record. The vast majority of soft-bodied forms, however, have left no record or have left a very incomplete fossil record. Second, the organism must be **buried rapidly** before the structures have a chance to decompose, erode, or disarticulate. Third, burial must effectively **seal the organism** from aerobic bacterial action and from decomposing chemical fluids.

Rapid burial, such as submergence in a peat bog, burial in sand or mud along a marine shore, trapping in pitch of conifer trees, or burial in ancient tar pools, would effectively seal organisms from bacterial action and preserve the hard parts. On the other hand, rapid burial in a broad alluvial fan or in a sandy area through which water can rapidly percolate is not likely to produce a fossil, because of chemical and biological decay, which will quickly set in and destroy any evidence of animal or plant.

Fossils are preserved by three major methods: **unaltered** soft or hard parts, **altered** hard parts, and **trace** fossils. Essentially unaltered organisms are rare in the geologic column and each time one is described it causes some excitement in scientific circles. Those fossils that are essentially unaltered have undergone little chemical or physical change since the death of the organism. Wooly mammoths and rhinoceroses trapped in glacial sediments in Siberia, Alaska, and the Yukon Territory are examples of unaltered preservation. To be frozen intact is a temporary type of fossilization, because a constant frigid environment is necessary. Such preservation is only temporary in terms of Earth's history.

Insects, spiders, and various other arthropods have been preserved in amber in Tertiary deposits along the Baltic Sea and in Mexico. The hard parts of these animals preserved under such circumstances must have remained unaltered for a long time although most of the soft parts were desiccated.

Unaltered hard parts have also been collected from ancient tar deposits, such as those at Rancho la Brea in Los Angeles, California, or in Argentina. At these sites, bones of Pleistocene birds, mammals, and some reptiles have been preserved and sealed from bacterial action by entombment in the tarry deposits. Because of the limited number of tar seeps, such fossilization is rare.

Much more common unaltered hard parts are the shells or tests of various invertebrate animals. Oysters and other mollusks have been preserved with the inner shell of "mother of pearl" layer still intact. Many conodonts, inarticulate brachiopods, and fragments of echinoderms have been preserved essentially unaltered because of the stable nature of their skeletal material. Often foraminifera and diatoms have their hard parts preserved unaltered in relatively young rocks. From the standpoint of total numbers of fossils, however, preservation of essentially unaltered materials is rare.

Another method of unaltered fossilization is **desiccation,** or driving off of water from tissues. This type of preservation is not common and results in rather temporary fossils in areas with an arid climate. An extinct mummified ground sloth was still articulated, and ligaments held the bones firmly together. Pieces of skin were still attached to the carcass in places, but in other areas the skin and muscles had been consumed by carnivores trapped in the same volcanic pit. Desiccated mummies of southwestern Indians are examples of the same type of preservation, although in this instance the organisms are very young geologically.

A second major method of fossilization is **alteration of hard parts.** In general, three types of alteration of the hard parts can be recognized: recrystallization, permineralization, and replacement. Figure 9.1 illustrates a hypothetical clam shell undergoing a variety of changes on its way to becoming a fossil.

In **recrystallization,** the original skeletal materials have been reorganized into different minerals or larger crystals of the same mineral. Aragonite, a skeletal material precipitated by many mollusks and other organisms, is a form of calcium carbonate that is unstable over long periods of time. Aragonite recrystallizes easily into the more stable form of calcium carbonate, called calcite. In recrystallization, no new material is added or taken away, but simply a rearrangement of the crystalline substances occurs. Recrystallization generally does not change the external form of the hard part but obscures or destroys internal structures precipitated by the organism.

Permineralization is another type of alteration and is particularly common in porous substances such as bone or wood. Permineralization is accomplished by the addition of materials to fill the pores of the structure and generally produces a very heavy bone or piece of wood. The skeletal structure may be in its original condition or it may have been replaced or recrystallized. Many dinosaur and mammal bones in Mesozoic and Cenozoic deposits of the West, and the much sought-after fossil wood of Mesozoic and Cenozoic deposits in many parts of the country, are a result of permineralization.

Replacement results in the removal of original skeletal material and subsequent replacement by a secondary compound. An example is original calcite or aragonite of invertebrate shells replaced by iron sulfide in the form of pyrite or marcasite. Silica is another common replacing material, and in instances where silica has replaced the shells in limestone, they are easily recovered from the surrounding rock by solution of the matrix in dilute hydrochloric acid. One of the most famous silicified faunas in North America is that of the Permian rocks of west Texas, where large numbers of organisms have been delicately silicified and preserved by replacement. Calcium carbonate may in turn replace structures that were originally siliceous. The original opaline spicules of fossil sponges are often replaced by coarse crystalline calcite if the matrix in which the sponges occur is calcareous. Limonite, gypsum, dolomite, native silver or copper, copper sulfate, and various phosphates have all been noted as substances that have replaced original material during fossilization. In replacement, permineralization, and recrystallization, a solid hard duplicate structure results from the chemical modification of the original hard parts. In only rare instances are soft parts replaced or permineralized as part of the fossilization process.

In many areas, particularly in humid regions, skeletal materials may be dissolved away, leaving a hollow impression in the rocks called a **mold.** This natural cavity may preserve, in varying detail, the shape of the original structure. A model of the outside of the shell is termed an **external mold,** whereas that of the inside is termed an **internal mold.** In some organisms, such as a clam, the inside area between the closed valves, or shells, may be filled with sediment, forming an internal mold of the clam shell. This sediment-filling of the inside of the clam shell is termed a **steinkern** and is a common method of preserving various mollusks such as clams and snails. If one fills the mold, either naturally or artificially, with a foreign substance, the foreign material then duplicates the shape of the original and is termed a cast. A **cast** is a replica of the original form. Natural casts occur where mineralizing solutions fill original molds by precipitation.

Trace fossils are of increasing interest to paleontologists. **Tracks, trails,** or **burrows** are commonly the only vestige of large populations of land or marine animals. Most tracks and trails, however, show us little of the real configuration of the organism, although they do tell us something of its size, weight, and perhaps even its feeding pattern. By their very nature, interpretation of tracks and trails is more problematical than interpreting most other types of fossils because of the lack of direct information concerning the shape of the original living organism.

Animal borings may also tell us something of the life habits of the organism. Because of the recent interest shown in interpretation of sedimentary sequence where tracks, trails, and borings are the only fossils preserved, the new field of **ichnology,** a study of indirect fossils, has developed within historical geology.

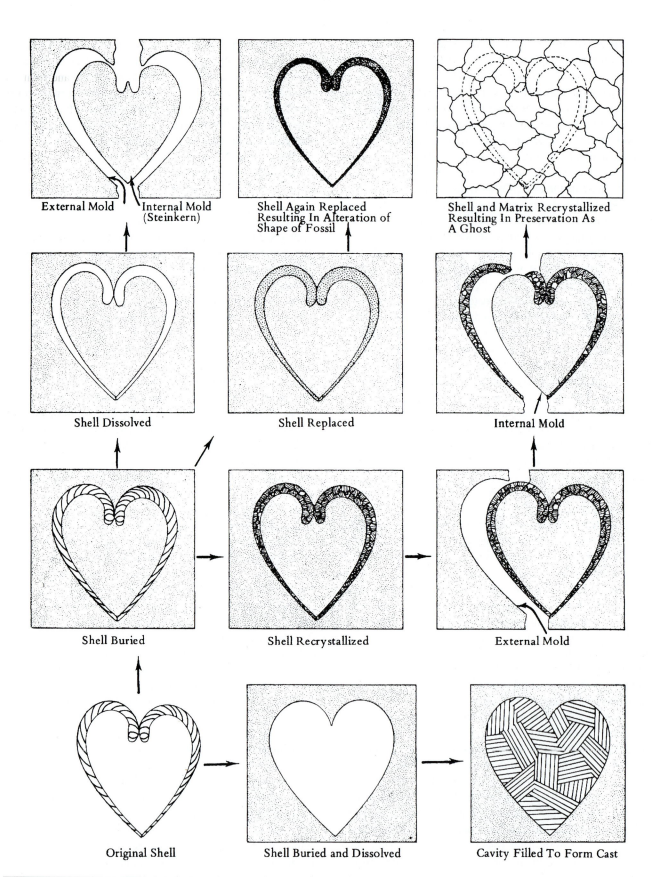

Figure 9.1 A diagram showing the various types of fossil preservation and their relationships. Uniform stipple pattern represents similar matrix. The arrows show possible sequences of preservation.

Carbonization, as a method of fossilization, occurs when the organism remains are preserved as a film of carbon. Such carbon films may show structures of the organism in great detail although they are now preserved only as a flat impression. Leaves, graptolites, crustaceans, and fish have been preserved in great numbers in limited areas by this method. One of the most famous carbonized faunas is that of the Middle Cambrian Burgess Shale, discovered near the turn of the century in British Columbia. Here carbonized films of even soft-bodied forms have been exquisitely preserved, so that details of the digestive tract, nervous system, and surficial hairlike bristles have been preserved as part of the fossil. Even though common under certain circumstances, carbonization is quantitatively a relatively minor method of fossilization.

Coprolites are solid excretory waste pellets of animals that are occasionally preserved. They may contain undigested material and thus give considerable information concerning the diet of the animal, its size, and where the animal may have lived. Coprolites may contain teeth, scales, plant remains, hard parts, or material that is not easily digested; as such they may be significant finds in terms of historical geology.

Gastroliths, or "stomach stones," have been described as associated with Mesozoic dinosaur remains throughout the world. These stones are highly polished and rounded, a result of grinding against one another as part of the fragmentation process in eating and digestion. Not all rounded, highly polished stones from even the most favorable horizons should be considered gastroliths, because many other processes can produce similar highly polished and rounded stones as well. Differentiation usually requires much experience, or an unusual association to clearly identify polished gastroliths. One of the most diagnostic features on some of the least questioned occurrences is an acidlike pitting in depressions beneath the general rounded, polished surface of the stone.

Procedure

1. Examine the various fossils prepared for the exercise and determine the methods of fossilization. Record the kinds of preservation, noting particularly those forms that are direct versus indirect fossils.

2. Summarize what can be told of the history of the organism or the history of its preservation as a fossil, using what can be seen in the adhering matrix as well as from the fossil.

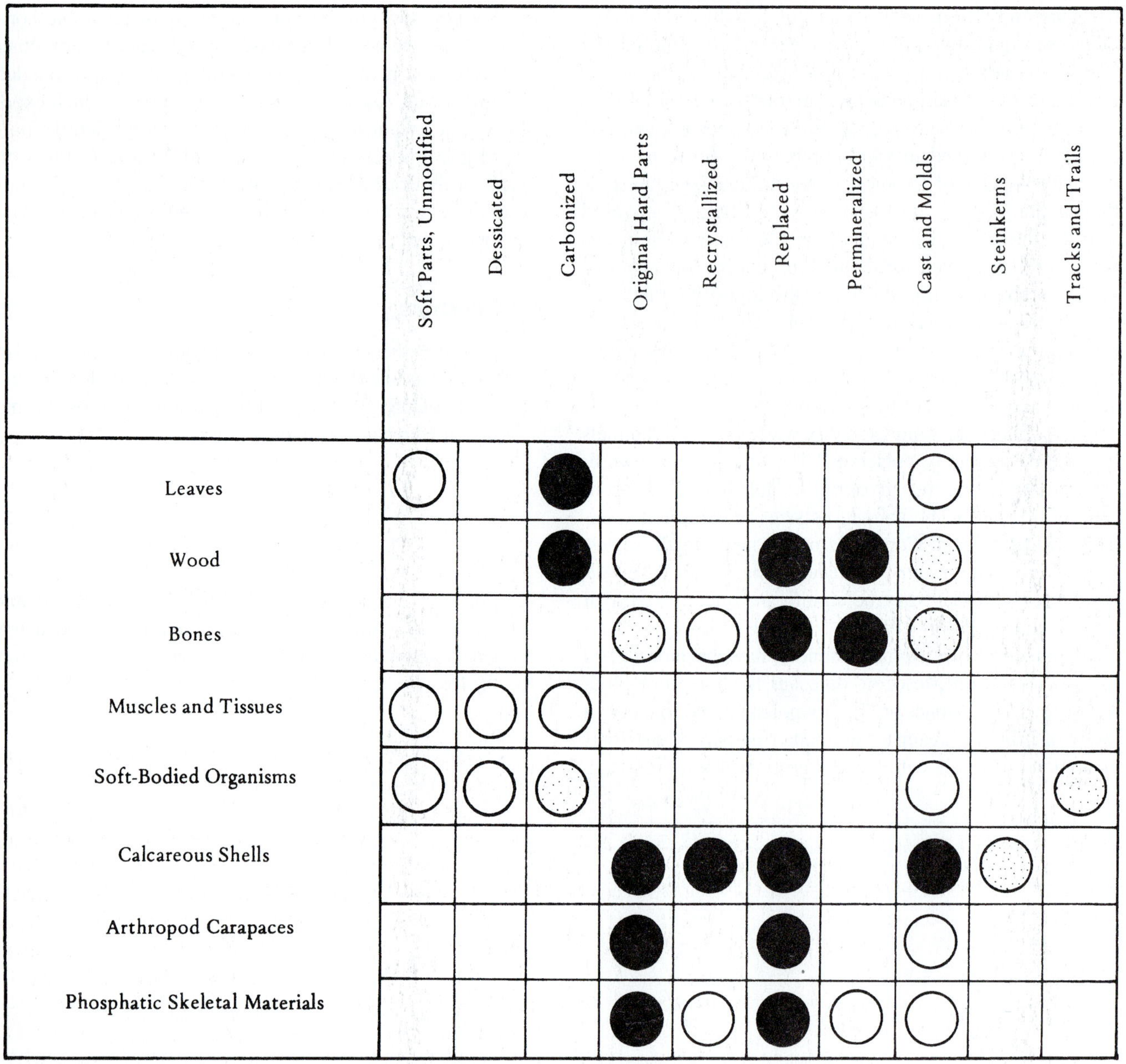

	Soft Parts, Unmodified	Dessicated	Carbonized	Original Hard Parts	Recrystallized	Replaced	Permineralized	Cast and Molds	Steinkerns	Tracks and Trails
Leaves	○		●					○		
Wood			●	○		●	●	⦿		
Bones				⦿	○	●	●	⦿		
Muscles and Tissues	○	○	○							
Soft-Bodied Organisms	○	○	⦿					○		⦿
Calcareous Shells				●	●	●		●	⦿	
Arthropod Carapaces				●		●		○		
Phosphatic Skeletal Materials				●	○	●	○	○		

Figure 9.2 A chart showing types of preservation characteristic of various organic materials. Solid black circles represent common methods, stippled circles denote less common methods, and open circles denote rare methods.

Fossil Classification and Morphology

Classification of fossil plants and animals is called **taxonomy.** The basis for taxonomy is similarity of morphology and shape, and phylogenetic relationships. In practice, all fossils are assigned two-part names, following what is called the **binomial system.** The first part of the name is the **generic** name; the second is the **specific** name. Occasionally a third or even a fourth part is used to show combinations of subgenera or subspecies. Ordinarily, only genus and species names are used. The Roman alphabet is used for all names in texts of any language. Conventionally, names are derived from Latin or Greek roots. The basis for subdivision and classification of organisms is the species. **Species** are defined as groups of organisms that normally interbreed and produce fertile offspring. This definition, although usable for neontologists or taxonomists of modern species, is not usable for the study of fossils in paleontology where the basis of classification is in fact morphology or similarity in form. Thus, we only infer in paleontology that morphological species were in fact interbreeding groups of organisms during their lifetime.

In table 10–1, the taxonomic hierarchies of a man and a dog are used as examples in the classification from kingdom to individual. Generic, subgeneric, specific, and subspecific names are italicized. Many of the scientific names are used as common words or in general ways with English plurals and nonitalicized forms. Table 10–2 provides an abbreviated classification of animals and plants.

Protozoans

Protozoans are unicellular organisms that range in size from microscopic to several millimeters in diameter. Protozoans live in marine environments as plankton and benthos, in fresh water, and as parasites in many living organisms.

Table 10–1	Hierarchy of Taxonomy	
Kingdom	Animalia	Animalia
Phylum	Vertebrata	Vertebrata
Class	Mammalia	Mammalia
Order	Carnivora	Primates
Family	Canidae	Hominidae
Genus	*Canis*	*Homo*
Species	*familiaris*	*sapiens*
Individual	Rover	John Brown

These small single cells perform all the living functions necessary for complete life cycles.

In spite of their size and relative simplicity, protozoans display a great variety of shapes and forms, and species are differentiated on that basis. Of the great variety of protozoans, only two groups that possess hard parts are important as fossils (fig. 10.1). The group called foraminifera are extremely abundant and are important in stratigraphy as time indicators. The foraminifera have shells or tests composed of calcium carbonate or of fine grains of minerals or rocks cemented together. The tests of forams are composed of chambers assembled in a variety of ways and patterns. One group of foraminifera, called fusulinids, built shells with about the same size and shape as grains of wheat or rice.

Another group of protozoans are radiolarians, a group that is abundant in modern seas and that has tests composed of concentric spheres or helmet-shaped, spiny structures of silica. Protozoans are known from rocks as old as Cambrian and even perhaps Precambrian, and they range to the Recent. Both foraminifera and radiolarians are abundant in the modern seas where their tests accumulate to form radiolarian and foraminiferal oozes. Fusulinids are used as

Table 10–2	Abbreviated Classification of Animals and Plants (as used in paleontology)

Animal Kingdom

Phylum	PROTOZOA
	Single cells, or groups of cells, generally microscopic foraminifers, radiolarians, fusulinids
Phylum	PORIFERA
	Sponges and stromatoporoids
Phylum	COELENTERATA
	Corals—tetracorals, hexacorals, and tabulate corals
Phylum	BRYOZOA
	Moss animals—small colonial animals
Phylum	BRACHIOPODA
	Bivalved invertebrates with unequal dorsal and ventral valves
Phylum	ECHINODERMATA
	Animals generally with fivefold radial symmetry, starfish, sand dollars, echinoids, sea lilies or crinoids, blastoids, cystoids
Phylum	MOLLUSCA
	Bivalves (Pelecypods—clam, oyster
	Gastropods—snail, slug
	Cephalopods—squid, octopus, nautiloid, ammonoid
Phylum	ANNELIDA
	Segmented worms
	Scolecodonts
Phylum	ARTHROPODA
	Invertebrate animals with jointed legs, insects, lobsters, crabs, trilobites, eurypterids
Phylum	HEMICHORDATA—Graptolites
Phylum	CONODONTA
Phylum	VERTEBRATA
	Animals with notochords and articulated backbones
	Pisces (fish)
	Amphibians
	Reptiles—dinosaurs, ichthyosaurs, plesiosaurs, mosasaurs
	Aves (birds)
	Mammals—warm-blooded animals, including humans

Plant Kingdom

Division Cyanophyta	blue-green algae
Division Chlorophyta	green algae
Division Phaeophyta	brown algae
Division Rhodophyta	red algae
Division Bryophyta	liverworts, hornworts, mosses
Division Psilophyta	psilophytes
Division Lycopodophyta	club mosses
Division Arthrophyta	horsetails
Division Pterophyta	ferns
Division Pteridospermophyta	seed ferns
Division Cycadophyta	cycads
Division Ginkophyta	ginkgos
Division Coniferphyta	confiers
Division Anthophyta	flowering plants
Class Dicotyledonae	dicots
Class Monocotyledonae	monocots

stratigraphic time indicators in Carboniferous and Permian rocks where they are abundant; other small foraminifera are used in the dating and correlation of Cretaceous and Tertiary rocks. Radiolaria are used as stratigraphic indicators for deep marine sediments. Because of their small size protozoans not only require identification and study with a microscope but also can be recovered in cuttings from drill holes of oil wells and are one of the most common and valuable fossils for stratigraphic studies in late Mesozoic and Cenozoic rocks.

Porifera

Porifera, or sponges, are simple multicellular animals that live attached to the substrate (figs. 10.2 and 10.3A). A sponge may be thought of as a vase-shaped animal. Its body walls are penetrated by a series of canals of varying degrees of complexity. Microscopic food particles are removed from currents of water that pass through the canal system of the sponge. Most sponges are marine but some occur in fresh water. The sponge is held more or less rigid by an internal stiffening skeleton of fibers of spongin or spicules made of calcium carbonate or silicon dioxide. The commercial sponge is the flexible skeleton (spongin) of a modern marine sponge from which all of the once-living protoplasm has been removed.

Because of the nature of the body of the sponge, complete animals are rarely preserved as fossils. The most common fossil remains are the individual minute spicules of silica or calcium carbonate that provided support for the flesh. The most complete fossil sponges are of groups where the skeleton was solidly fused during life. These sponges appear as conical, spherical, and platterlike fossils that are perforated by numerous small canals.

A group of sponges important as fossils are the stromatoporoids (fig. 10.3A). They were a colonial group of organisms that secreted a calcareous laminated skeleton. These laminated colonial skeletal structures resemble heads of cabbage or flat-lying sheets. Other more unusual stromatoporoids formed twiglike structures with a dendritic or branching mode of growth. Stromatoporoids are found in rocks ranging in age from Cambrian to Cretaceous and are quantitatively important in rocks of Ordovician and Devonian ages.

Phylum Coelenterata

Within the phylum Coelenterata, two groups are important as fossils. The first group, the corals, are among the most abundant fossils in sedimentary rocks, and are typically found as either "horn corals," the skeletal remains of a single organism (fig. 10.3B), or as a group of individual skeletons cemented together to form a colony (fig. 10.3 C–D). Solitary corals have a pit in which the animal was attached to the top, broad part of the calcareous "horn." The "horn"

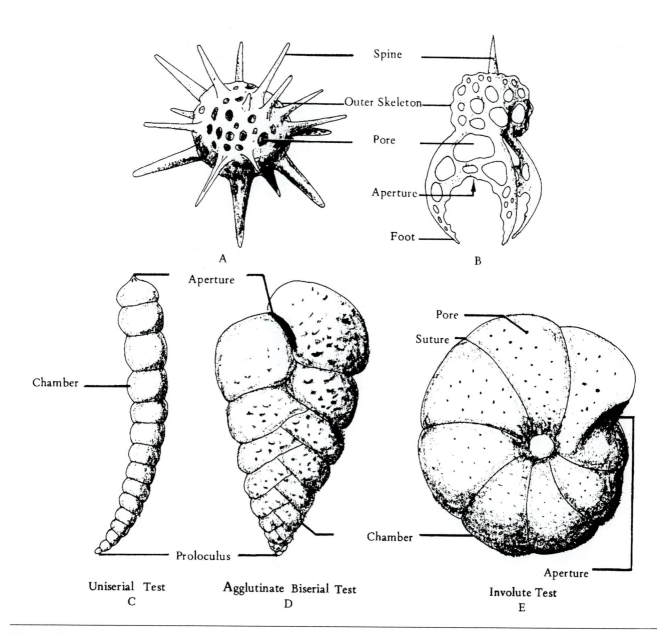

Figure 10.1 Morphology of radiolarians and foraminifera. **A–B,** radiolarians (×100); **C–E,** foraminifera (×100).

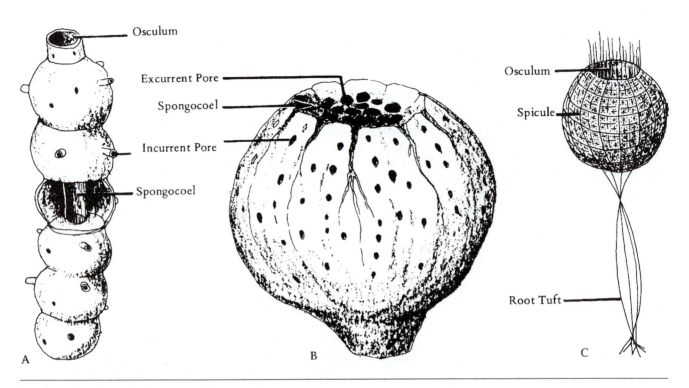

Figure 10.2 Morphology of fossil sponges.
A. calcareous sponge; **B.** lithistid sponge;
C. hexactinellid sponge.

functioned as an external skeleton and was built in daily increments by the organism. Corals first appeared in Ordovician rocks; continued as a major group through the Paleozoic, Mesozoic, and Cenozoic; and are common in modern seas. Corals form reefs in modern seas and inhabit the shallow areas of the ocean near the equator, and it is presumed that they did so in the past.

Phylum Bryozoa

Bryozoans, the moss animals, are small aquatic organisms that secrete colonial calcareous external skeletons and are common as fossils (fig. 10.4). Most bryozoans were marine but a few freshwater forms are also known. They are more advanced than corals and have a nervous system and a U-shaped digestive tract. The colonial organisms lived in the minute pores or tubes that perforate the stony skeletal structure. Fossil bryozoans resemble bits of lace or small twiglike structures, often encrusting other organisms or fossils. They are usually preserved lying parallel to the bedding planes of the enclosing layers. Bryozoa are found in rocks that range in age from the Cambrian to the Recent and are

particularly abundant in rocks of Mississippian to Permian ages. Because of their small size, relatively slow evolution, and difficulty of identification, fossil bryozoans have not been used extensively in stratigraphic determinations.

Brachiopods

Brachiopods are marine invertebrates that were much more abundant in the seas of the Paleozoic Era than they are today (fig. 10.5). They range in size from less than an inch to approximately 6 inches at their broadest point. Their calcareous or chitinophosphatic shells consist of two unequal valves that are symmetrical when divided into lateral halves. This shell shape distinguishes the brachiopods from the bivalves (clams, etc.), which are equivalved with right and left valves that are essentially mirror images of each other (fig. 10.6). They are among the most abundant fossil types found in rocks of the Paleozoic Era. Their shells are preserved in nearly every type of sedimentary rock. Because of their abundance, their great variety in shell form, and ease of identification, brachiopods are extremely useful as time and ecologic indicators in stratigraphic studies.

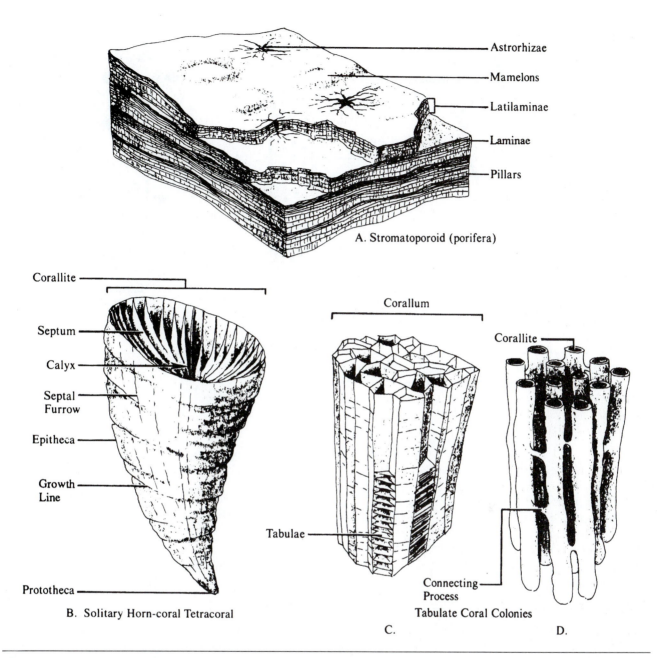

A. Stromatoporoid (porifera)

Astrorhizae

Mamelons

Latilaminae

Laminae

Pillars

Corallite

Septum

Calyx

Septal Furrow

Epitheca

Growth Line

Prototheca

B. Solitary Horn-coral Tetracoral

Corallum

Corallite

Tabulae

Connecting Process

Tabulate Coral Colonies

C.

D.

Figure 10.3 Morphology of porifera **(A)** and coelenterates **(B–D). A.** Stromatoporoid; **B.** Tetracoral; **C–D.** Colonial tabulate corals.

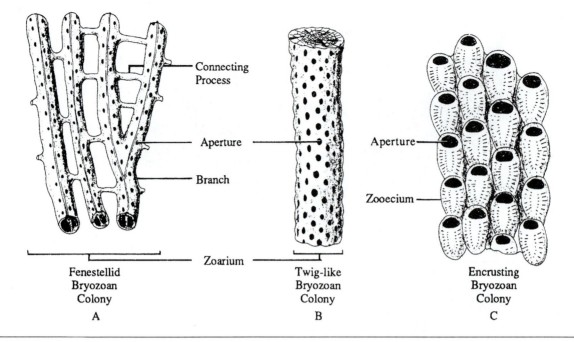

Figure 10.4 Morphology of bryozoans. **A.** Fenestellid type (×3). **B.** Twiglike type (×5). **C.** Encrusting cheilostome type (×5).

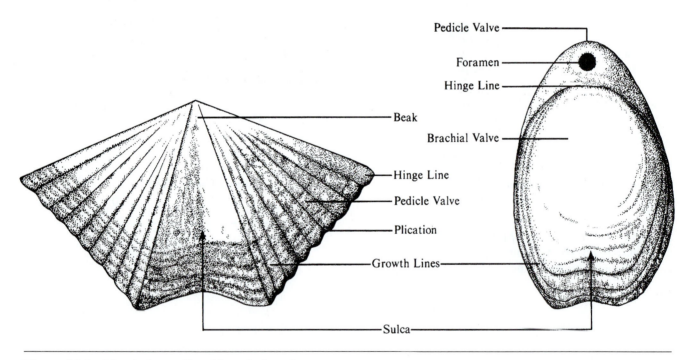

Figure 10.5 Morphology of brachiopods, based upon two common types.

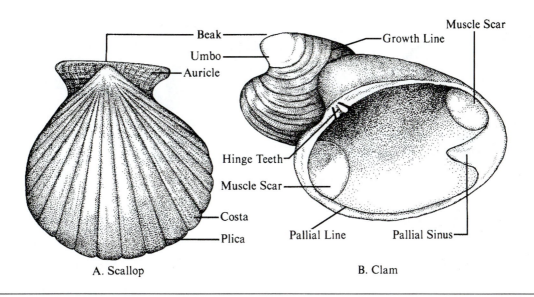

Figure 10.6 Morphology of bivalves. **A.** Scallop.
B. Clam.

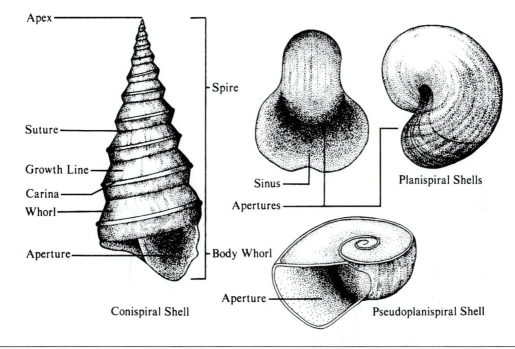

Figure 10.7 Morphology of gastropods, showing
three common conch shapes.

Mollusca

The phylum Mollusca includes animals that are zoologically similar, but superficially different because of the great diversity in the shapes of their shells. Included in the phylum are the bivalves (clams, scallops, and oysters), gastropods (snails, slugs, pteropods) (fig. 10.7), cephalopods (*Nautilus,* squid, octopus, and cuttle fish) (fig. 10.8), and other less common forms such as the chitons and scaphopods.

As fossils, mollusks are common in marine and non-marine rocks from the Cambrian to the Recent. Their calcareous shells are valuable indicators of time and ecology. One group of cephalopods, the ammonoids (fig. 10.8A), have complexly chambered shells and are used as a world-wide standard of reference in biostratigraphy from the Devonian through the Cretaceous. Mollusks have adapted at present to many ecologic niches from deep ocean benthonic environments to air-breathing existence on mountain peaks at 18,000 feet above sea level.

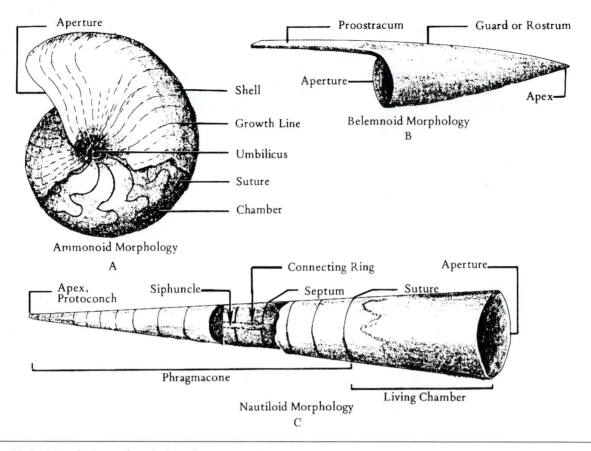

Figure 10.8 Morphology of cephalopods.
A. Ammonoid; **B.** Belemnite; **C.** Nautiloid.

Arthropods

The arthropods are characterized by their chitinoid, segmented exoskeleton, or carapace. This group includes insects, crustaceans (crabs, lobsters, etc.), chelicerates (spiders, etc.), myriapods (centipedes and millipedes), and trilobites. Of these groups, only the trilobites and a type of crustacean, the ostracodes, are common as fossils (fig. 10.9). Trilobites especially are abundant fossils in Cambrian, Ordovician, and Silurian rocks. Because arthropods shed their chitinous exoskeletons along joints, their fossil remains often consist of disarticulated heads, tails, or body segments.

Both trilobites and ostracodes are important fossils as time indicators. Trilobites lived from the Cambrian through the Permian, and the ostracodes lived from the Cambrian to the Recent. Trilobites are particularly useful guide fossils for Cambrian and Ordovician rocks, and ostracodes have been most used in rocks from the Ordovician, Silurian, Devonian, and Cenozoic.

Echinoderms

Echinoderms, or "spiny-skinned animals," are exclusively marine organisms and include the modern starfish, brittle stars, sand dollars, sea urchins, sea cucumbers, and sea lilies. This group of organisms was much more common in the geologic past than it is at present. Echinoderms typically display fivefold radial symmetry. Their skeletons are formed of calcite plates, which are secreted inside an outer tissue layer, and form an external skeletal covering. Echinoderms that are important as fossils include crinoids, blastoids, cystoids, and echinoids (fig. 10.10). Echinoids are found throughout the entire geologic column from Lower Cambrian to Recent; however, they are more typical of Carboniferous, Permian, and Tertiary deposits. Blastoids, cystoids, and crinoids all reached high points of evolution during the Paleozoic. Many limestones of Mississippian age contain abundant crinoid stem fragments, sometimes making up the bulk of the rock mass.

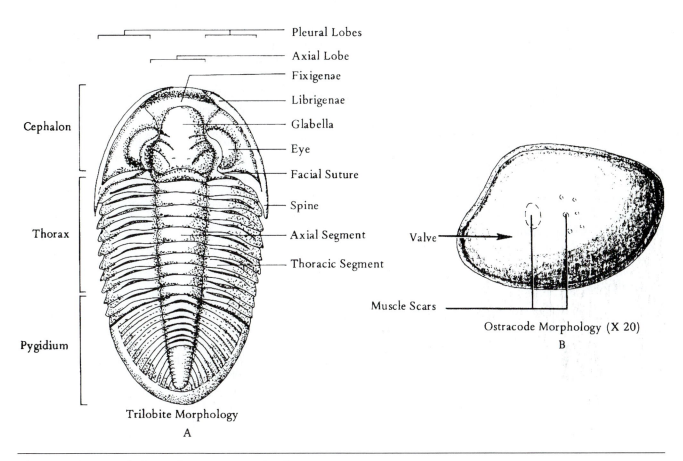

Trilobite Morphology
A

Ostracode Morphology (X 20)
B

Figure 10.9 Morphology of two common fossil arthropods. **A.** Trilobite; **B.** Ostracode.

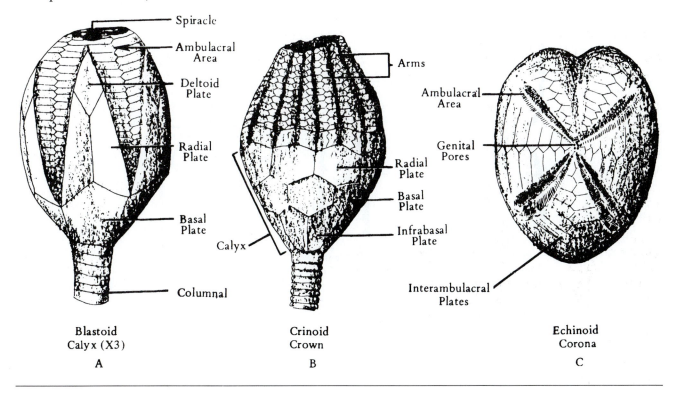

Blastoid
Calyx (X3)
A

Crinoid
Crown
B

Echinoid
Corona
C

Figure 10.10 Morphology of three common fossil echinoderms. **A.** Blastoid (×3); **B.** Crinoid; **C.** Irregular echinoid.

Graptolites

Graptolites are a group of extinct colonial organisms whose proteinacious, saw bladelike remains appear superficially similar to pencil marks on the surface of rocks, thus their name, which means "writing on rocks." They are typically small fossils; colonies ordinarily measure a few centimeters across. The individuals of a colony are nearly microscopic in size (fig. 10.11).

Graptolites are most important as time indicators during the Ordovician, Silurian, and Devonian periods. The Ordovician Period is often called the Age of Graptolites. Some graptolites lived into the Mississippian, but the group is not useful for time determination after the lower part of the Devonian.

Conodonts

Conodonts are an extinct group of phosphatic microfossils whose zoological affinities were unknown until recently. They are now placed in a separate phylum, Conodonta. Conodonts are found as disarticulated elements (fig. 10.12). Elements are parts of an apparatus within the conodont animal (fig. 10.13). Conodont elements are generally designated P, M, and S according to their presumed positions within the apparatus. A conodont species contains representative P, M, and S elements within its apparatus. The contained elements as well as the resulting apparatus are unique for any given taxon. Generic and specific names of

conodonts refer to the makeup of the apparatus, whereas the separate elements represent only a disarticulated portion of the remains of the conodont organism. Conodont elements have either right- or left-hand symmetry. Each symmetrical type is positioned on the appropriate side of the medial line of symmetry (fig. 10.13).

Most conodonts are less than 1 mm in size and are medium to dark brown in color. They can be found in almost all kinds of sedimentary rock from the Cambrian to the Triassic. In the Ordovician, Devonian, Mississippian, and Triassic periods, they are the most useful fossils available for intercontinental correlation because of their widespread occurrence, abundance, and highly predictable evolutionary patterns.

Procedure

The information presented in this exercise is provided to acquaint the student with the important morphologic features of the major fossil groups. After studying actual specimens provided by your instructor, identify the morphology that is preserved on your study materials. Indicate, by marking on the diagrams, those morphologic features you have actually observed in your study. It is common for a single specimen to display some morphologic elements, while others are not preserved. Observe several specimens of each type of fossil until you have personally observed all important elements of morphology.

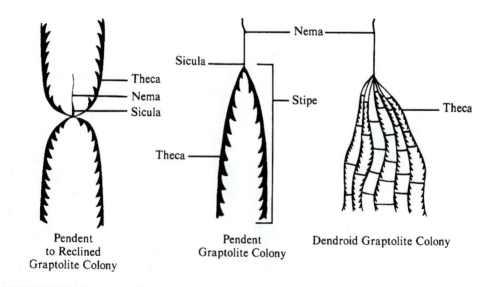

Figure 10.11   Morphology of three common types of graptolites (×2).

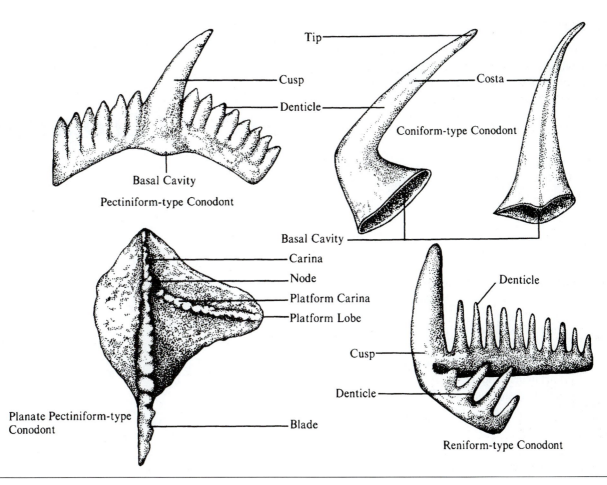

Figure 10.12 Morphology of conodont elements, showing parts of four common types (×50).

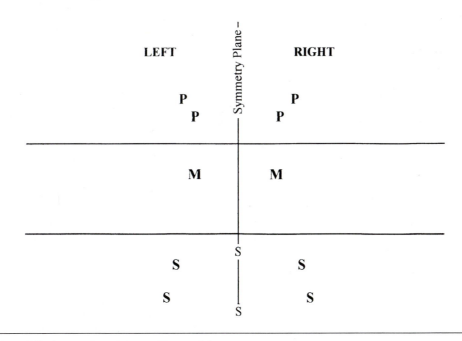

Figure 10.13 Simplified map showing positions of the three types of conodont elements, P, M, and S within an apparatus.

Evidence of Evolution

In only a few centuries, by selective breeding, humankind has produced an amazing array of variation in several species, for example, dogs, cats, horses, fruits, and cereal grains. This human-made change through time is much more rapid, but otherwise is very similar to nature's change through time—**evolution.**

The results of evolution are evident upon close examination of living as well as fossilized organisms. Through the study of fossils, at least three lines of evidence have contributed to our understanding of evolution: vestigial structures, ontogenetic change, and homologous bone structure.

Vestigial structures are organs or features that at one time were functional but have now lost their usefulness. For example, the pineal body in the human is a remnant of a third eye that is still well developed in some reptiles. A small bony tail and ear-moving muscles are present in humans although their functions have been lost, and the small toe shown in figure 11.5G was a vestigial structure in three-toed horses. These organs are vestiges of once functional features.

Ontogeny describes the growth or development of individuals throughout their life spans. Some organisms retain all growth stages in their development from immature to mature growth forms. One such group is the ammonoids, whose growing shell encompasses the previously deposited part of the shell to form an expanding planispiral structure. By examining the growth stages of ammonoids, as well as some other forms, it was seen that the development of an individual resembles the evolutionary stages in the history of the individual's lineage, or **phylogeny.**

Procedure

Part A

In figure 11.1, the ontogeny of an ammonoid genus, *Perrinites* of Permian age, is illustrated. The sutures, A–F, represent growth stages of the individual, A being the earliest, and F representing the mature stage. In figure 11.2, only the mature sutures of six ammonoid genera from Pennsyl-

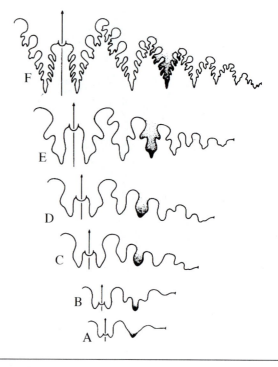

Figure 11.1 Drawings of sutures showing the ontogenetic development of the Permian ammonoid genus *Perrinites.* (From *Treatise on Invertebrate Paleontology,* courtesy of the Geological Society of America and the University of Kansas.)

vanian and Permian rocks are illustrated. These genera are not contemporaneous, but are found in rock units of various ages, although within these two periods.

1. The stratigraphic positions of the ammonoids, as represented by their sutures, have intentionally been rearranged. By studying the ontogeny of *Perrinites* (fig. 11.1), rearrange the sutures in figure 11.2 into proper stratigraphic order, based on the biogenetic law, and compare with the order given by your instructor.

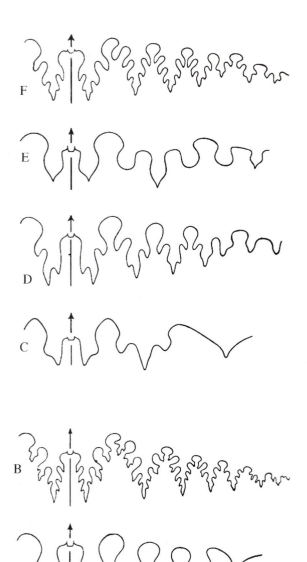

Figure 11.2 Drawings of adult sutures of six ammonoids, in a random arrangement of ages. **A.** *Shumardites.* **B.** *Perrinites.* **C.** *Aktubites.* **D.** *Properrinites.* **E.** *Parashumardites.* **F.** *Metaperrinites.* (From Ruzhencev.)

Procedure

Part B

Skeletal elements of structures that appear similar in different groups of organisms are termed **homologous structures,** even though their functions and overall shapes change slightly from group to group. For example, the arm and hand of a human are similar to, or have homologous bones with, the front limb of a horse, the wing of a bat or bird, or the front limb of a marine reptile. The relative position and structural relationships of the respective bones remain the same, but various types of limbs can be produced by changing the shape of individual bones. This similarity is the result of modification through time to accommodate a particular environment or organic function.

1. Various bones are identified on the skeleton of a man in figure 11.3. Homologize as many of these bones as possible on the skeleton of *Dimetrodon* (fig. 11.4). *Dimetrodon* is one of the fin-backed, mammal-like reptiles of the Late Paleozoic.

2. Color each of the homologous bones of the front limb of the mammals, reptiles, and birds illustrated in figure 11.5. Color the scapula green, the humerus red, the radius blue, the ulna pink, the metacarpals yellow, and the phalanges purple.

3. Summarize the modifications that are evident in the various limbs and explain, if possible, the reason for nature's selection of the modification, in view of the environment where the various organisms lived.

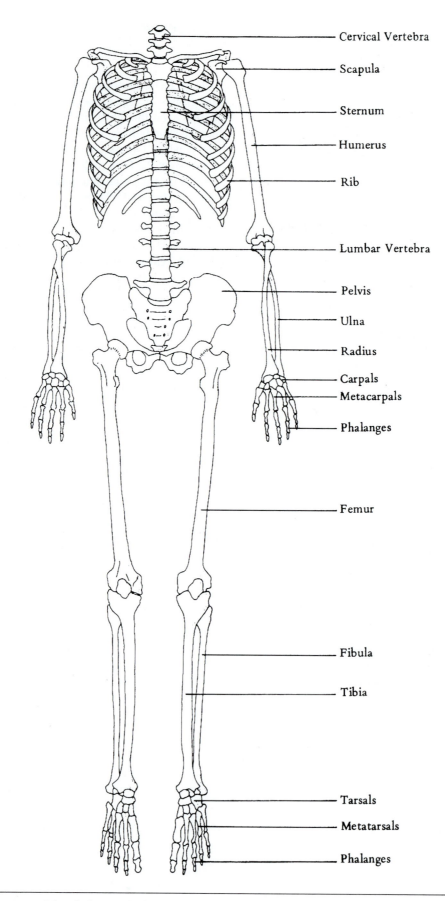

Figure 11.3 Diagram of the skeleton of a human, with various bones labeled.

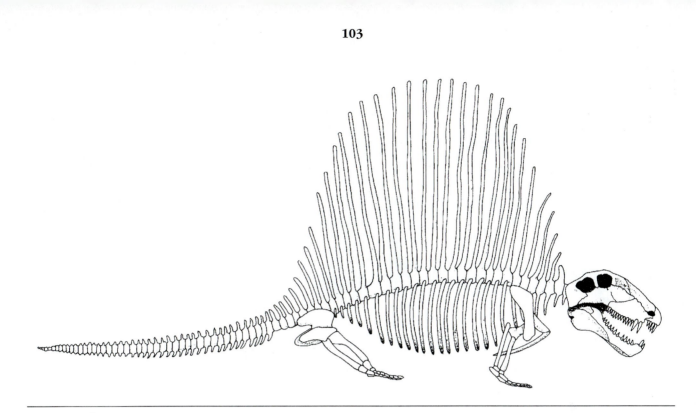

Figure 11.4 Diagram of a skeleton of the fin-backed
Permian reptile, *Dimetrodon.*

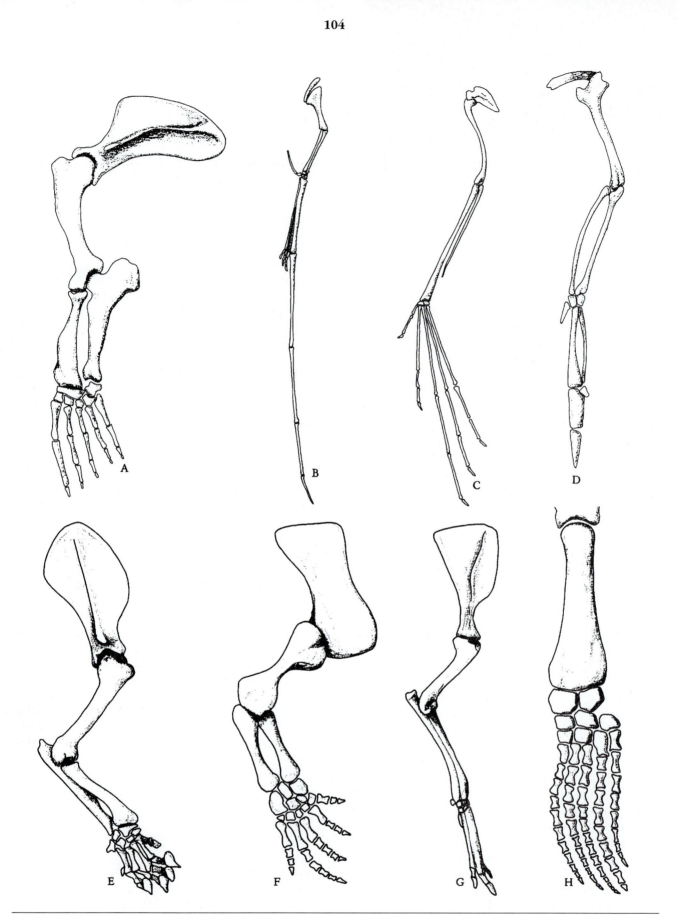

Figure 11.5 Diagrams of the forelimbs of eight vertebrate animals. **A.** Modern seal. **B.** Cretaceous pterodactyl. **C.** Modern bat. **D.** Modern bird. **E.** Quaternary sabre-toothed cat. **F.** Permian mammal-like reptile, edaphosaur. **G.** Oligocene three-toed horse. **H.** Jurassic plesiosaur.

Patterns of Evolution

The fossil record is one of the most conclusive pieces of evidence of change in organisms through geologic time. Since the first appearance of life on Earth, sometime during the Precambrian, organisms have undergone continual change, or evolution. Lines of descent of organisms through time are termed **phylogenies,** and it is by a study of fossils that phylogenetic or evolutionary patterns of various groups can be reconstructed. Studies of living organisms demonstrate relationships in terms of protein kinships, structural similarities, vestigial organs, and various other types of evidences, but these are circumstantial, and inconclusive at best. The geologic record documents the sequence of forms as they existed on Earth and preserves these phylogenetic patterns of evolution.

Generally speaking, the soft parts of fossil organisms are not available for study, because only the hard parts are preserved. This preservation naturally limits the interpretive value of fossils in reconstruction of lineages because many of the most characteristic features of organisms are lost in the process of fossilization. The aspect of the organism that remains is usually only the form or shape of the original hard parts. A phylogeny of fossil material is, therefore, an interpretation based on comparison of similar forms, their stratigraphic occurrence, and apparent changes in morphology. Such studies are called **comparative morphology.**

Various patterns of evolution, or inferred phylogenies, are illustrated by fossil study. Some of these patterns are illustrated on figure 12.1. Convergence is one pattern commonly followed by evolving organisms. **Convergence** is a pattern of evolution that is the result of dissimilar or unrelated organisms converging or becoming more similar-appearing through time, probably in response to life in similar environments. If the convergence is essentially contemporaneous in both lineages, the pattern is termed **isochronous** convergence. An example is shown in figure 12.1D. If the convergence, however, occurs at different times, such as illustrated in figure 12.1E, the pattern is termed **heterochronous** convergence. Several remarkable examples of convergence are found in fossil records (for

example, the development of the marsupial Tasmanian wolf of the Australian region to that of the placental doglike forms in the northern hemisphere). The convergence in gross shape of mammalian dolphins and whales to that of rapidly swimming fish is immediately apparent. Several fossil reptiles also show the same general trend toward streamlining. The latter instance is an example of heterochronous convergence since the swimming reptiles were at their peak of development in the Mesozoic and the mammals at their peak in the later Cenozoic. The spectacular development of flight in reptiles (pterodactyls), mammals, and birds is another example of convergence because of the adaptation of each group to a similar mode of life. Examples can also be drawn from among invertebrate organisms. Various genera of brachiopods and bivalves converge at different times to form structures similar to solitary horn corals. Mimicry in insects might also be considered examples of convergent evolution where similarity to the successful group is protective.

Parallel evolution is another pattern that is demonstrated in the fossil record where two lineages show the same general trend in morphology. In dinosaurs, for example, there is a general tendency in several lineages for an increase in size. Fusulinid foraminifera show other examples of parallelism where several lineages show a gradual increase in wall thickness and a corresponding increase in complexity of wall folding and other structural elements. These fossils occur in the same beds and were undoubtedly contemporaneous. In ammonoid mollusks, the same patterns may be seen where separate lineages appear to independently develop increasingly complex shell chamber walls.

Divergence is one of the most common evolutionary patterns in the geologic record. Divergence occurs where two closely related lineages in the geologic record become less and less similar as they are traced in younger and younger rocks. Morphologic differences are interpreted to suggest change in environment, with one lineage specializing in one environment and the other in a distinctively different one. An example of divergence can be seen in

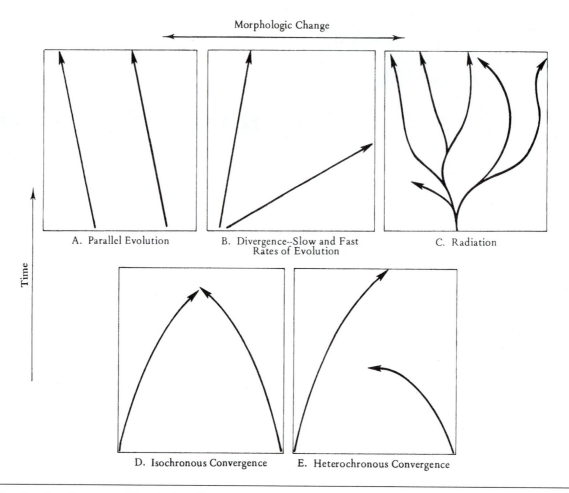

Figure 12.1 Graphs of patterns of evolution. Time is plotted vertically and morphologic change is plotted horizontally.

horses, where browsing and grazing horses diverge from one another in the middle part of the Tertiary Period. The illustration of divergence, figure 12.1B, shows relative rates of evolution with a slow-rate and a fast-rate line. The slow-rate line illustrates the least change over the longest period of time and is the steeper of the two orthogenetic trends shown in figure 12.1B. The fast-rate line has the more gentle gradient and the greatest amount of structural change in a short time. Examples of very rapid evolution can be seen in many groups of invertebrate fossils and in some of the more specialized vertebrate fossils as well, where considerable morphologic change was accomplished in a very short time. Ammonoids, some mammals, and graptolites all have high-rate evolutionary lineages. Slow evolutionary rates are exemplified by some inarticulate brachiopods, which have changed little since the group appeared in the Cambrian over 500 million years ago. Other classic examples are forms like the opossum, which has changed little since the early Tertiary.

Radiation is a common pattern and is seen where several lineages diverge from one another. An example of radiation is seen in the Mesozoic history of reptiles, which radi-

ated to become the largest herbivores and carnivores on land. Some reptiles began to fly, and others returned secondarily to the sea. Several lineages of reptiles and mammals have returned to the sea; however, they have not developed gills but have retained lungs like their terrestrial ancestors.

Not since the years directly following Darwin's monumental publication has paleontology more strongly influenced evolutionary thought than in the last 15 years. Happily, during the past decade, the fossil record, now much better understood than in the late nineteenth century, has become a fertile field of data to test modern ideas. For the past several years, two ideas have been popular points of debate among paleontologists and neontologists. The ideas, called **phyletic gradualism** and **punctuated equilibria,** are models for the mode and tempo of evolution. They do not challenge the idea of evolution, only the mode and tempo of the theory. The basic question debated was, Does evolution proceed gradually in a continuous pattern, or is it episodic?

If gradualistic, then the fossil record should display intermediate forms as progenitors, responding to a slowly changing environment, giving rise to descendants. The two

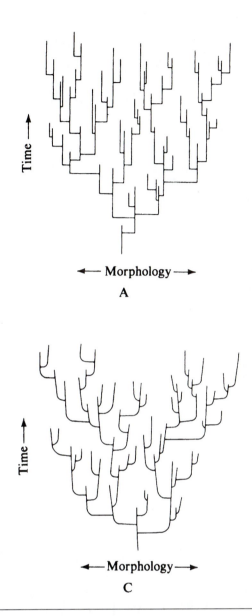

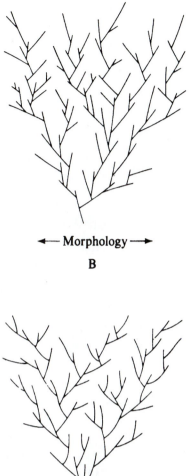

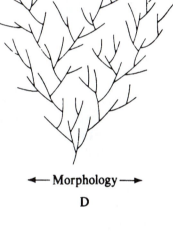

Figure 12.2 Hypothetical phylogenies representing: **A.** Extreme view of punctuational model. Sudden change in morphology (evolution) followed by static condition. **B.** Extreme view of gradualistic model. Evolution is gradual, continuous, and nearly uniform as shown by slope of lines representing species. **C.** Punctuationalist view with some gradualistic influence. **D.** Gradualistic model with some accommodation for punctuationism. (After Stanley, S. M., 1979, *Macroevolution*, W. H. Freeman.)

species would show such a complete transition in their morphology that separating the two would be arbitrary. On the other hand, if episodic, the fossils should show abrupt change between species followed by a relatively long period of stasis (fig. 12.2A). Proponents of punctuated equilibria were successful in discovering numerous examples from the fossil record to support their interpretation. Supporters of the gradualistic model also found support from some fossil groups, especially mammals and protozoans (fig. 12.2B). Most students today see evidence for both (fig. 12.2C, D).

Procedure

Part A

The drawings in figure 12.3 illustrate geometric forms from eight successive stratigraphic horizons. The oldest, at the base, represents the lowermost series of beds in the sequence and contains the most primitive shapes. Younger beds successively yield newer or more specialized kinds of fossils, until the youngest fossils are encountered in the topmost bed of the sequence.

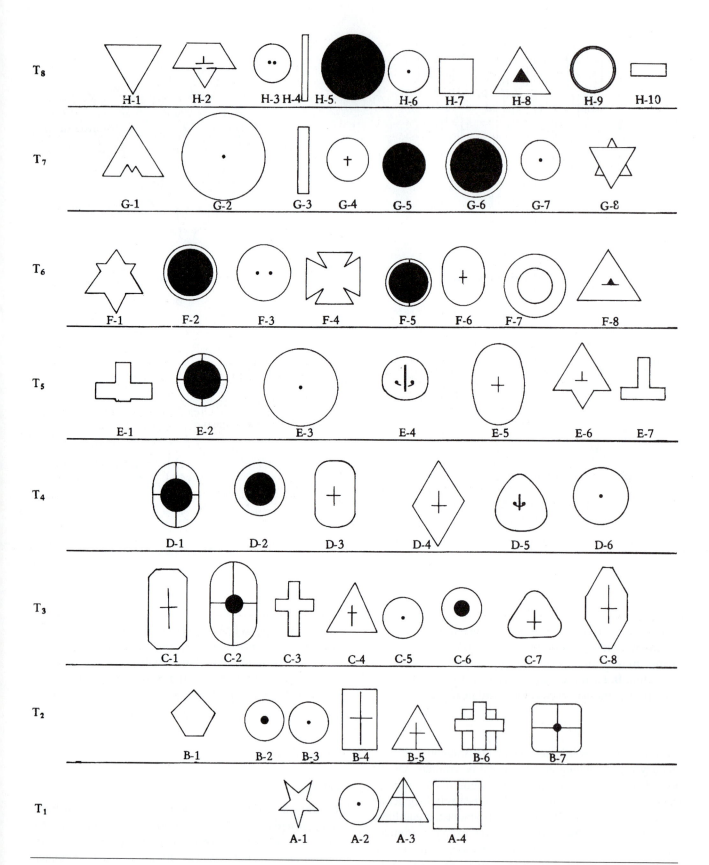

Figure 12.3 A diagram of various geometric shapes to simulate kinds of fossils in eight stratigraphic intervals, to be used in a problem showing evolutionary patterns and phylogenetic trends. Time is represented by the letter "T;" the oldest horizon is labeled T_1.

Because many students have trouble manipulating complex shapes, such as those characteristic of many fossil lineages, the shapes here have been simplified to give only a sequence of geometric forms. These forms were constructed to show various kinds of evolutionary patterns, such as convergence, parallel evolution, divergence, and radiation. In several instances, there is no single correct or right answer. There are some forms that—by convergence—may be produced as a result of either of two or more converging lineages. These are termed **polyphyletic** forms. Other lineages seem best interpreted as deriving from a single root stock, which is called **monophyletic.**

1. Begin with the oldest series of shapes (layer A) and work upward, connecting by pencil lines or identifying by letter and number, the likely phylogenetic series in each successive layer.
2. Tear out (or copy) the page and cut the fossils apart, rearranging them into patterns somewhat like that of the graph of radiation in figure 12.1C. You may identify the fossils by letter and number and report your work.

3. Identify by number and letter an example of a convergent, parallel, divergent, and radiating trend in the evolutionary pattern. Cite examples of isochronous as well as heterochronous convergence, if such occur. Can you see monophyletic and polyphyletic lines?

4. Connect the various series with arrows to show trends. Indicate with asterisks those lineages that became extinct, based on the horizon or level of occurrence of the last form in the series.
5. Is the dominant mode and tempo of evolution in the relationships shown in figure 12.3 gradualistic or punctuational?

Part B

Figure 12.4 illustrates 10 stratigraphic horizons of graptolite faunas, with the oldest at the bottom and the youngest at the top. The entire sequence of organisms in figure 12.4 represents fossils that first appeared in the Upper Cambrian or Lower Ordovician rocks and continued through the Silurian into the Devonian. Each of the collections is distinctive of a separate faunal zone of early Paleozoic graptolite evolution. Graptolites are an extinct group whose characteristics evolved rapidly through geologic time. Because of their rapid change, they are used as time stratigraphic indicators in Ordovician and Silurian deposits.

1. Establish possible lineages or phylogenies for the various graptolites shown, as was done on figure 12.3, with a series of pencil lines showing phylogenies on the page. A more clearly defined interpretation can be obtained by copying the page and cutting the various forms apart, rearranging them in likely series. Be careful to maintain the fossils at their correct stratigraphic horizon. Each of the graptolite zones represents several million years of time and therefore, it is possible to have one form derived from another within the same stratigraphic interval. In many instances, the entire suite of fossils illustrated may occur on a single bedding plane and may record an instant in terms of organic evolution.
2. Locate with numbers and letters possible examples of parallel evolution, convergence, divergence, and a possible polyphyletic origin.

Part C

Figure 12.5 illustrates mammal teeth in a stratigraphic succession like the graptolites in figure 12.4. By comparing the kinds of teeth shown in the illustration, reconstruct possible evolutionary trends in this group of vertebrates. Teeth serve well for such an exercise, because the tooth structures of each mammal are highly specialized and are a major basis of classification in this group of organisms.

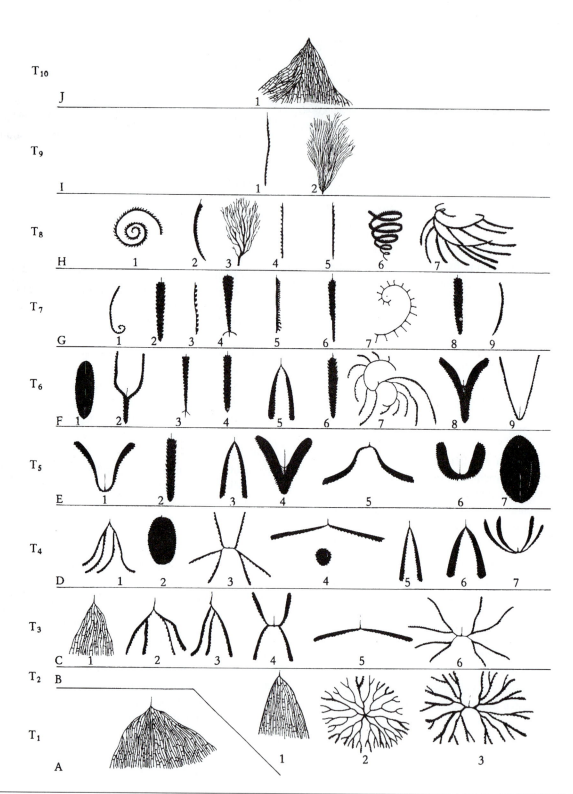

Figure 12.4 A diagram of graptolite occurrences from ten successive stratigraphic horizons, a problem in reconstruction of phylogenic trends. Time is represented by the letter "T;" the oldest horizon is T_1.

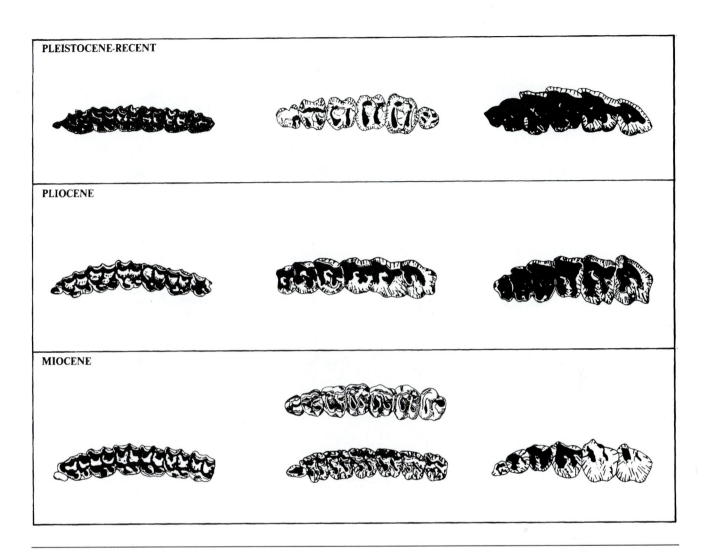

Figure 12.5 A sampling of mammal cheek teeth (upper) as they occur in Cenozoic strata. The evolutionary trends can best be seen by studying the patterns of the tooth surfaces and the general outline of the individual teeth. The dark areas on the teeth are areas of exposed dentine due to wear. The basic pattern of this dentine is a distinctive characteristic of every mammal, and an excellent means of identification. (Drawings after Hatcher, 1896; Matthew, 1915; Osborn, 1929; Cook, 1930; Scott, 1937; Gregory, 1951; Pivetau, 1958; and Romer, 1966.)

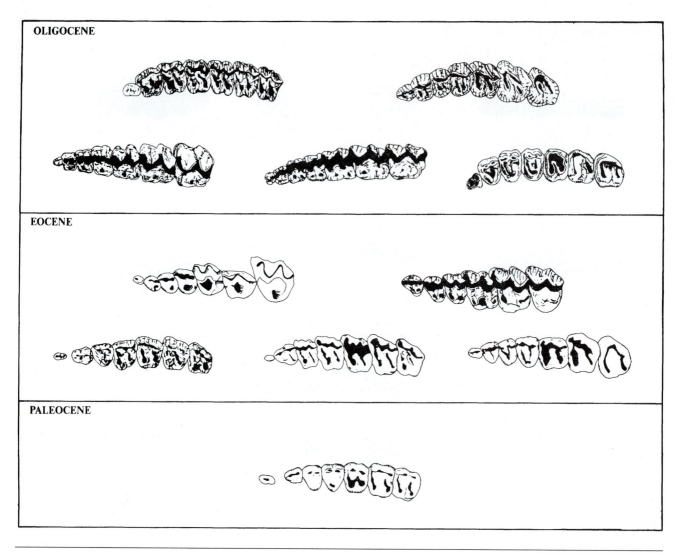

Figure 12.5 continued

Biostratigraphic Correlation

ossils are the practical "clocks" of geology. Correlation by the use of fossils to determine contemporaneity of rock units is **biostratigraphic correlation** or **biostratigraphy.**

William Smith, an English surveyor, discovered in 1816 that various kinds of fossils appear in an orderly and predictable fashion in a vertical succession of rock units. Based on Smith's findings, the **principle of faunal succession** was formulated. The principle states that each successive time interval of Earth history contains a unique assemblage of fossil kinds and therefore rock units can be dated by analysis of their fossil content and comparison to standard reference sequences. Because fossils occur almost exclusively in sedimentary rocks, biostratigraphic correlation is ordinarily not possible in igneous or metamorphic rocks. When clear relationships of igneous rocks are lacking and when radiometric dates are unavailable, time correlation of these rocks is problematical.

Although nearly all kinds of fossils are used for geologic dating on a local scale, only a select few groups are utilized for intercontinental and regional correlation as reference standards in the geologic column. In figure 13.1, the relative value and use for various groups of fossils as reliable time indicators is summarized.

Fossils that are best suited for time correlation over great distances, such as between continents, should have the following characteristics: abundance (they must be readily available in order to be of much use), wide geographic range (their usefulness is in direct proportion to the area throughout which they are found), rapid evolutionary pattern (this results in a short geologic time range), rapid dispersion (this minimizes differences of age in different parts of the world), and easy identification (if only a few experts can distinguish one from another, their value is compromised).

Fossils of this type are normally of planktonic or nektonic organisms. Organisms possessing all of these characteristics are rare, but planktonic (floating) or nektonic (swimming) organisms are generally best. A floating or free-swimming life habit allows rapid distribution in many environments.

Examples of various fossils shown in figure 13.1 that are useful for time stratigraphic determinations include trilobites in the Cambrian; fusulinid foraminifers in the Pennsylvanian and Permian systems; smaller foraminifera in the Tertiary; graptolites of the Ordovician, Silurian, and earliest Devonian; ammonoids of the Devonian through the Cretaceous; and conodonts of the Silurian, Devonian, Mississippian, and Triassic systems.

In actual practice, use of fossils in determining age of rocks employs the use of a concept called a **zone.** A zone is a subdivision of a **stage.** It represents a restricted sequence of rock layers, or strata, that represent the physical record of a discrete interval of Earth history. It is identified by the presence of a specified group of fossil taxa, one of which serves as a name bearer. For example, the *Cheiloceras* Zone is present when strata yield species of Cheiloceras, and/or other Late Devonian ammonoids that are characteristic of that time interval in the physical record. Some fossil zones are extremely limited in their geologic time range and their presence discriminates a small segment of the geologic column and thus a limited time interval. Within the Jurassic, for example, approximately 60 ammonoid zones are recognized whose individual time duration is thought to be approximately one-half million years. Conodonts of the Late Devonian allow zonation to perhaps as little as 300,000 years, which is almost instantaneous geologically speaking. The origin of the zone concept in biostratigraphy has persisted with remarkably little modification since its original description by a young German paleontologist, Albert Oppel, in 1856–1858.

Procedure

Part A

1. Using the data presented in figure 13.2A, complete the chart shown in figure 13.2B by marking the stratigraphic range of each of the remaining nine species, using species A as an example.

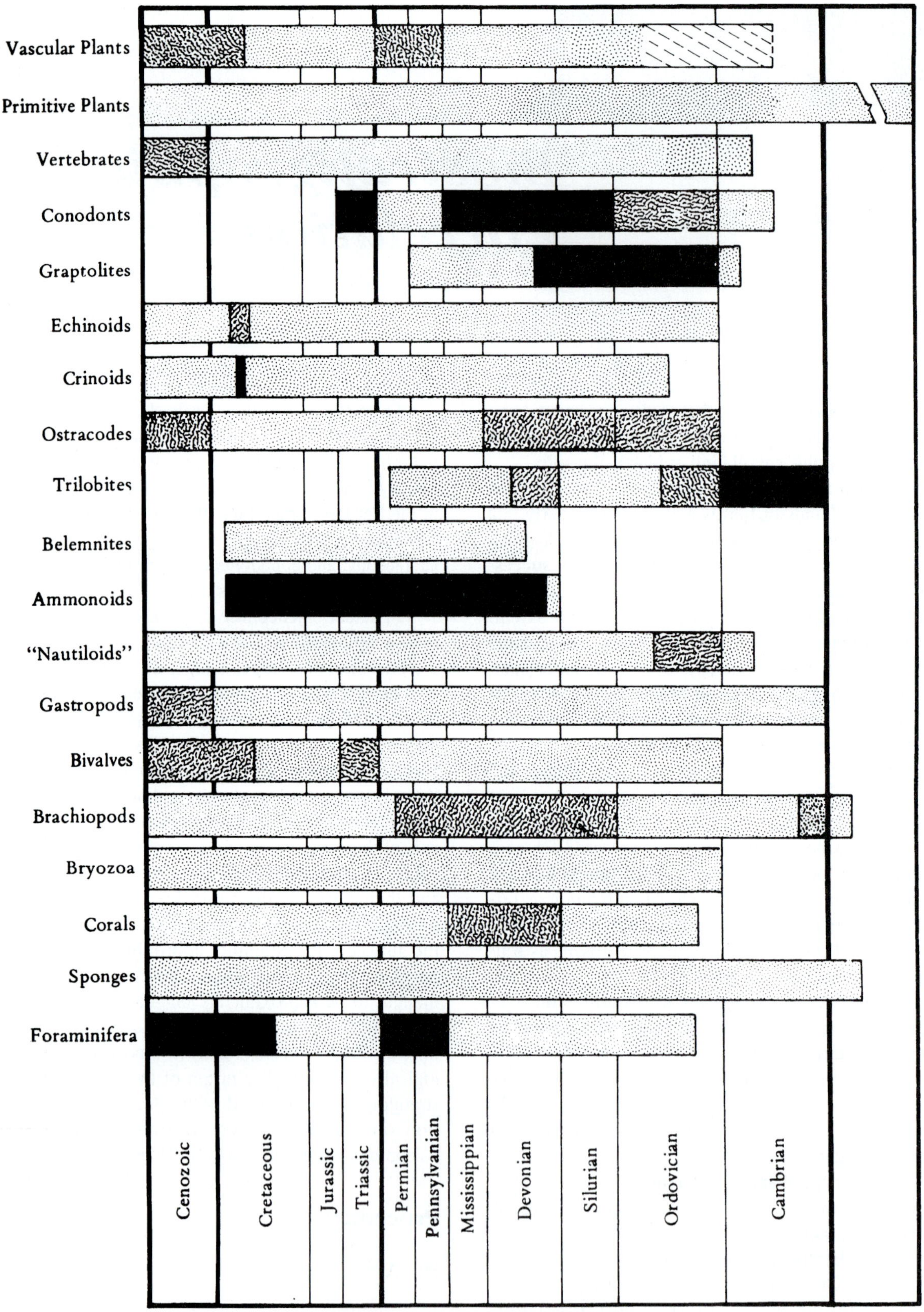

Figure 13.1 A chart showing the relative usefulness of various fossil groups as time indicators throughout the geologic column. The black columns indicate common use in worldwide correlation; dark stippled columns indicate local use or limited use in worldwide correlation; and light stippled columns indicate rare use in correlation; dashed diagonal lines indicate uncertainty. (After Teichert, 1958.)

Occurrence in m / Species	20	40	60	80	100	120	140	160	180	200
Species A	X	X	X							
Species B			X	X	X	X				
Species C			X	X						
Species D			X	X	X	X				
Species E			X	X	X	X	X			
Species F			X	X	X	X				
Species G					X	X	X	X		
Species H						X				
Species I							X	X	X	
Species J								X	X	X

A

Species / Thick in m	Species A	Species B	Species C	Species D	Species E	Species F	Species G	Species H	Species I	Species J	Zone
200											
180											
160											
140											
120											
100											
80											
60											
40											
20											

B

Figure 13.2 A. Table recording the occurrence of 10 species of fossils in a hypothetical sequence of rocks. **B.** Chart to be completed by the student illustrating stratigraphic ranges of species and their proposed zonation. The range of species A has been plotted as an example.

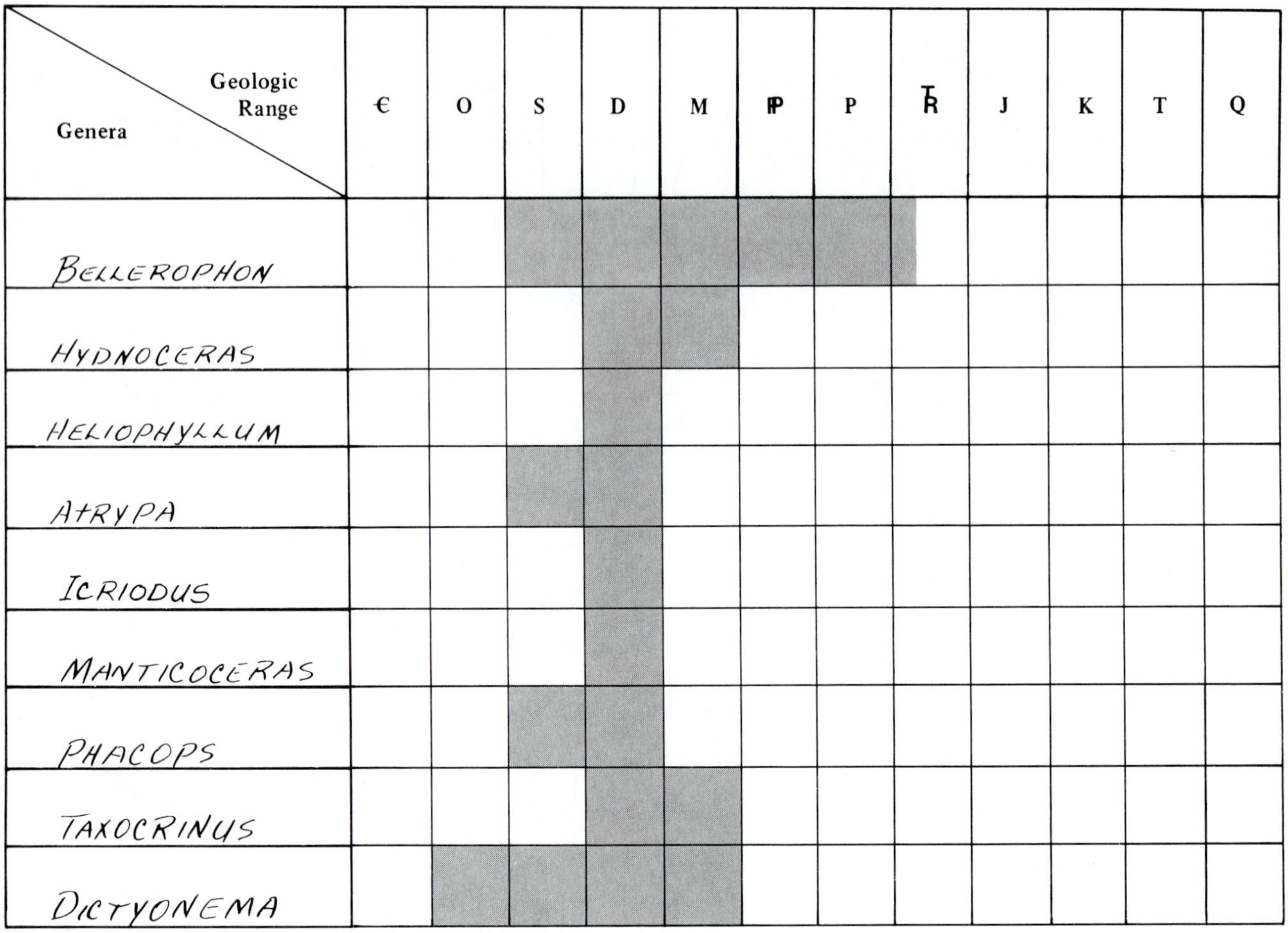

Genera \ Geologic Range	€	O	S	D	M	ℙ	P	Ꞃ	J	K	T	Q
BELLEROPHON			■	■	■	■	■					
HYDNOCERAS				■	■							
HELIOPHYLLUM				■								
ATRYPA			■	■								
ICRIODUS				■								
MANTICOCERAS				■								
PHACOPS			■	■								
TAXOCRINUS				■	■							
DICTYONEMA		■	■	■	■							

Figure 13.3 A chart showing the procedure to be followed in exercises on biostratigraphic dating. The association illustrated is representative of a Devonian fauna as indicated by the overlapping ranges of genera in the collection.

2. Subdivide the rocks by identifying possible zones on the basis of the fossil occurrence. Pick a name for each of your zones, for example Zone A, Zone 3, or the like, using individual species as the name bearer.

3. What is the average number of species occurring in each zone?

4. Give an example of a zone in which the name bearer never occurs outside its zone.

5. Give an example of a zone in which the name bearer occurs in another zone as well.

Part B

1. Compare the fossils supplied in the laboratory sets with the illustrations on the 13 plates that follow. Identify each of the fossils in the set with its generic name, and record the generic name on the worksheets on pages 133–140, as in figure 13.3, along with the geologic range of the genus. The range of each genus is given with each illustration. The age of the fauna illustrated in figure 13.3 is Devonian.

FORAMINIFERA

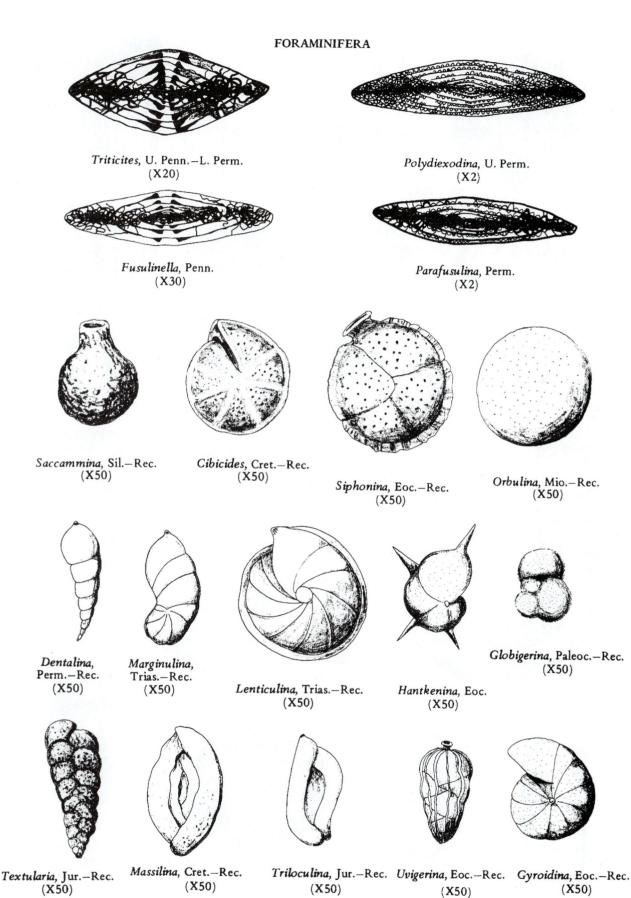

Triticites, U. Penn.—L. Perm.
(X20)

Polydiexodina, U. Perm.
(X2)

Fusulinella, Penn.
(X30)

Parafusulina, Perm.
(X2)

Saccammina, Sil.—Rec.
(X50)

Cibicides, Cret.—Rec.
(X50)

Siphonina, Eoc.—Rec.
(X50)

Orbulina, Mio.—Rec.
(X50)

Dentalina,
Perm.—Rec.
(X50)

Marginulina,
Trias.—Rec.
(X50)

Lenticulina, Trias.—Rec.
(X50)

Hantkenina, Eoc.
(X50)

Globigerina, Paleoc.—Rec.
(X50)

Textularia, Jur.—Rec.
(X50)

Massilina, Cret.—Rec.
(X50)

Triloculina, Jur.—Rec.
(X50)

Uvigerina, Eoc.—Rec.
(X50)

Gyroidina, Eoc.—Rec.
(X50)

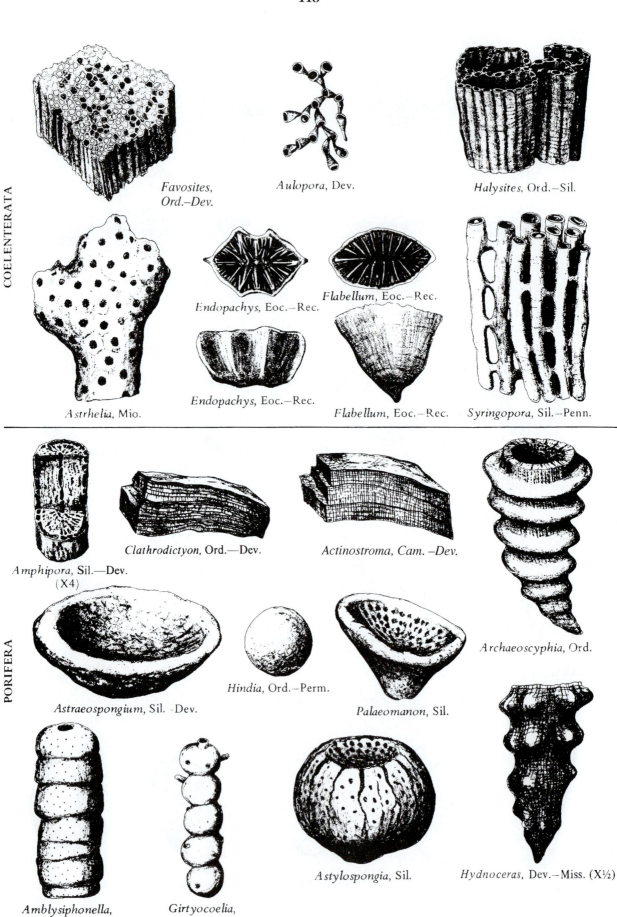

COELENTERATA

Favosites, Ord.–Dev.

Aulopora, Dev.

Halysites, Ord.–Sil.

Astrhelia, Mio.

Endopachys, Eoc.–Rec.

Endopachys, Eoc.–Rec.

Flabellum, Eoc.–Rec.

Flabellum, Eoc.–Rec.

Syringopora, Sil.–Penn.

PORIFERA

Amphipora, Sil.—Dev. (X4)

Clathrodictyon, Ord.—Dev.

Actinostroma, Cam. –Dev.

Archaeoscyphia, Ord.

Astraeospongium, Sil. -Dev.

Hindia, Ord.–Perm.

Palaeomanon, Sil.

Amblysiphonella, Penn.–Perm.

Girtyocoelia, Penn.–Perm.

Astylospongia, Sil.

Hydnoceras, Dev.–Miss. (X½)

COELENTERATA

Streptelasma, M. Ord.–M. Sil.

Caninia, Miss.–Perm.

Heliophyllum, Dev.

Zaphrenthis, Dev.

Heliophyllum, Dev.

Streptelasma, M. Ord.–M. Sil.

Caninia, Miss.–Perm.

Zaphrenthis, Dev.

Microcyclus, M. Dev.

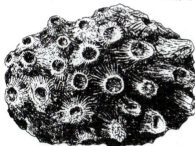

Cystiphyllum, Sil.

Pachyphyllum, U. Dev.

Calceola,
L. Dev.–M. Dev.

Lophophyllidium,
Penn.–Perm.

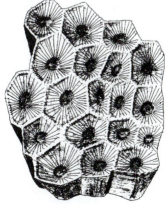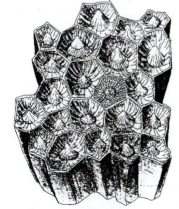

Eridophyllum, L. Dev.–M. Dev.

Hexagonaria, Dev.

Lithostrotionella, Miss.–L. Perm.

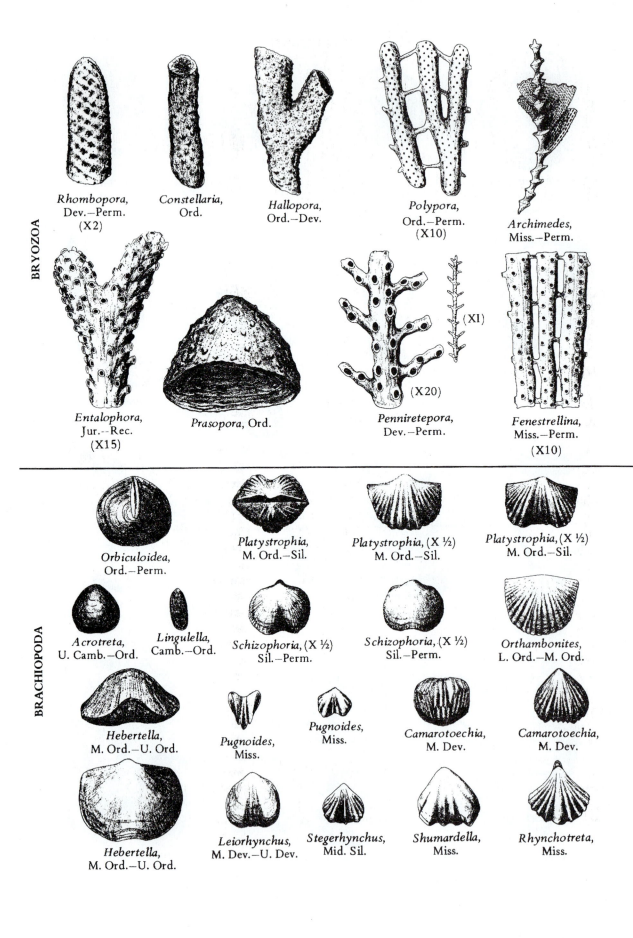

BRYOZOA

Rhombopora,
Dev.–Perm.
(X2)

Constellaria,
Ord.

Hallopora,
Ord.–Dev.

Polypora,
Ord.–Perm.
(X10)

Archimedes,
Miss.–Perm.

Entalophora,
Jur.--Rec.
(X15)

Prasopora, Ord.

Penniretepora,
Dev.–Perm.
(X20)

(XI)

Fenestrellina,
Miss.–Perm.
(X10)

BRACHIOPODA

Orbiculoidea,
Ord.–Perm.

Platystrophia,
M. Ord.–Sil.

Platystrophia, (X ½)
M. Ord.–Sil.

Platystrophia, (X ½)
M. Ord.–Sil.

Acrotreta,
U. Camb.–Ord.

Lingulella,
Camb.–Ord.

Schizophoria, (X ½)
Sil.–Perm.

Schizophoria, (X ½)
Sil.–Perm.

Orthambonites,
L. Ord.–M. Ord.

Hebertella,
M. Ord.–U. Ord.

Pugnoides,
Miss.

Pugnoides,
Miss.

Camarotoechia,
M. Dev.

Camarotoechia,
M. Dev.

Hebertella,
M. Ord.–U. Ord.

Leiorhynchus,
M. Dev.–U. Dev.

Stegerhynchus,
Mid. Sil.

Shumardella,
Miss.

Rhynchotreta,
Miss.

BRACHIOPODA

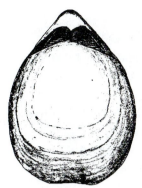

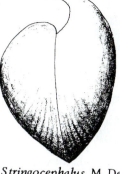

Stringocephalus, M. Dev.

Stringocephalus, M. Dev.

Mesolobus,
M. Penn.—L. Perm.

Chonetes,
Sil.—Perm.

Penicularis L. Perm.

Kingena, Cret.

Athyris, Dev.—Trias.

Juresania,
Penn.—L. Perm.

Juresania,
Penn. – L. Perm.

Composita,
U. Dev.—Perm.

Dielasma,
U. Miss.—Perm.

Dielasma,
U. Miss.—Perm.

Rafinesquina,
M. Ord.—U. Ord.

Rafinesquina,
M. Ord.—U. Ord.

Composita,
U. Dev.—Perm.

Hustedia,
Miss.—Perm.

Derbyia,
Miss.—Perm.

Derbyia,
Miss.—Perm.

Strophomena,
M. Ord.—U. Ord.

Strophomena,
M. Ord.—U. Ord.

Leptaena, M. Ord. – Dev.

Terebratula, Cret.—Tert.

Rensselaeria, L. Dev.

Echinoconchus, Miss.

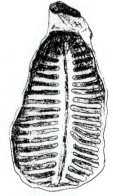

Leptodus, Perm.

BRACHIOPODA

Paraspirifer, L. Dev.–M. Dev.

Paraspirifer, L. Dev.–M. Dev.

Paraspirifer, L. Dev.–M. Dev.

Cyrtospirifer, U. Dev.–L. Miss.

Cyrtospirifer, U. Dev.–L. Miss.

Cyrtospirifer, U. Dev.–L. Miss.

Spirifer, Miss.–M. Penn.

Spirifer, Miss.–M. Penn.

Punctospirifer, Miss.–Perm.

Punctospirifer, Miss.–Perm.

Mucrospirifer, M. Dev.

Mucrospirifer, M. Dev.

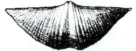

Mucrospirifer, M. Dev.

Cyrtina, Sil.–Perm.

Cyrtina, Sil.–Perm.

Atrypa, Sil.–Dev.

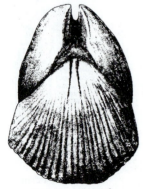

Pentamerus, Sil.

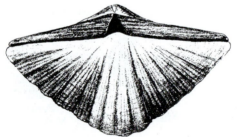

Neospirifer, Penn.–Perm.

Atrypa, Sil.–Dev.

Pentamerus, Sil.

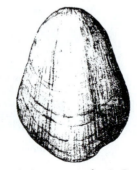

Conchidium, U. Ord.–L. Dev.

Conchidium, U. Ord.–L. Dev.

BIVALVIA (PELECYPODA)

Exogyra, Jur.–Cret. (X ½)

Trigonia, Jur.–Rec.(X ½)

Glycimeris, Cret.–Rec.

Arctica, Jur.–Rec.

Monotis, Trias.

Arca, Jur.–Rec.

Nuculana, Sil.–Rec.

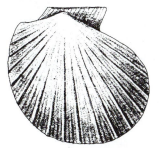

Lyriopecten, Dev.

Camptonectes, Jur.–Cret.

Myalina, Miss.–Perm.

Wilkingia, Miss.—Perm.

Gryphaea, Jur.–Eoc.

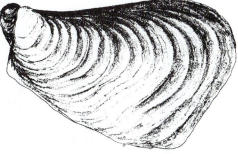

Inoceramus, Jur.–Cret.

Unio, Trias.–Rec.

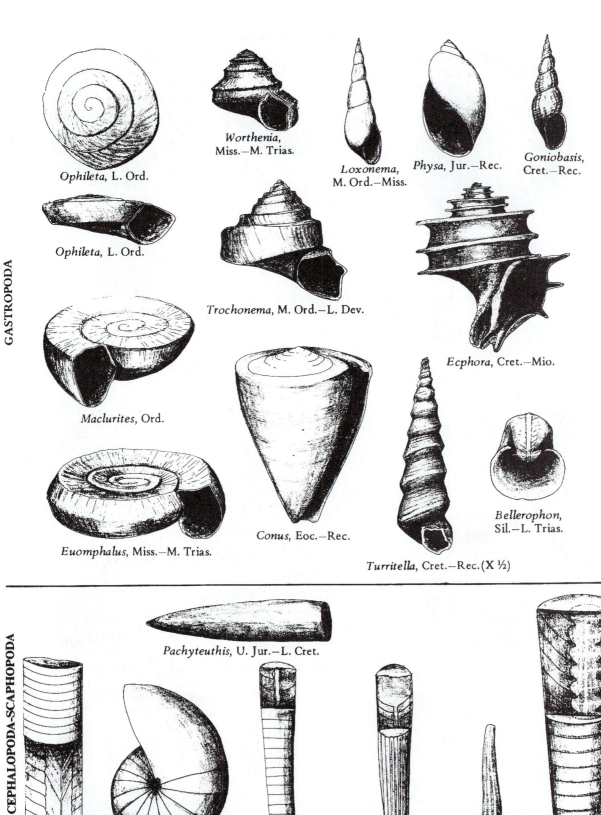

GASTROPODA

Ophileta, L. Ord.

Worthenia,
Miss.–M. Trias.

Loxonema,
M. Ord.–Miss.

Physa, Jur.–Rec.

Goniobasis,
Cret.–Rec.

Ophileta, L. Ord.

Trochonema, M. Ord.–L. Dev.

Ecphora, Cret.–Mio.

Maclurites, Ord.

Conus, Eoc.–Rec.

Bellerophon,
Sil.–L. Trias.

Euomphalus, Miss.–M. Trias.

Turritella, Cret.–Rec. (X ½)

CEPHALOPODA-SCAPHOPODA

Pachyteuthis, U. Jur.–L. Cret.

Endoceras, (X ½)
M. Ord.–U. Ord.

Eutrephoceras, (X ½)
U. Jur.–Mio.

Mooreoceras,
U. Dev.–L. Perm.

Kionoceras,
M. Ord.–L. Perm.

Dentalium,
M. Trias.–Rec.
(Scaphopoda)

Rayonnoceras,
Miss.–Penn.

AMMONOIDEA

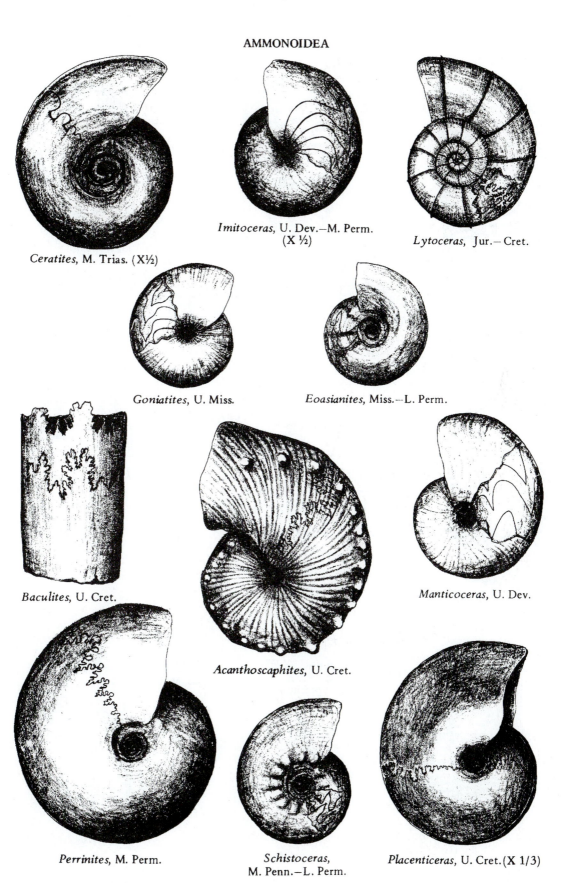

Ceratites, M. Trias. (X½)

Imitoceras, U. Dev.–M. Perm.
(X ½)

Lytoceras, Jur.– Cret.

Goniatites, U. Miss.

Eoasianites, Miss.–L. Perm.

Baculites, U. Cret.

Acanthoscaphites, U. Cret.

Manticoceras, U. Dev.

Perrinites, M. Perm.

Schistoceras,
M. Penn.–L. Perm.

Placenticeras, U. Cret.(X 1/3)

TRILOBITA—EURYPTERIDA

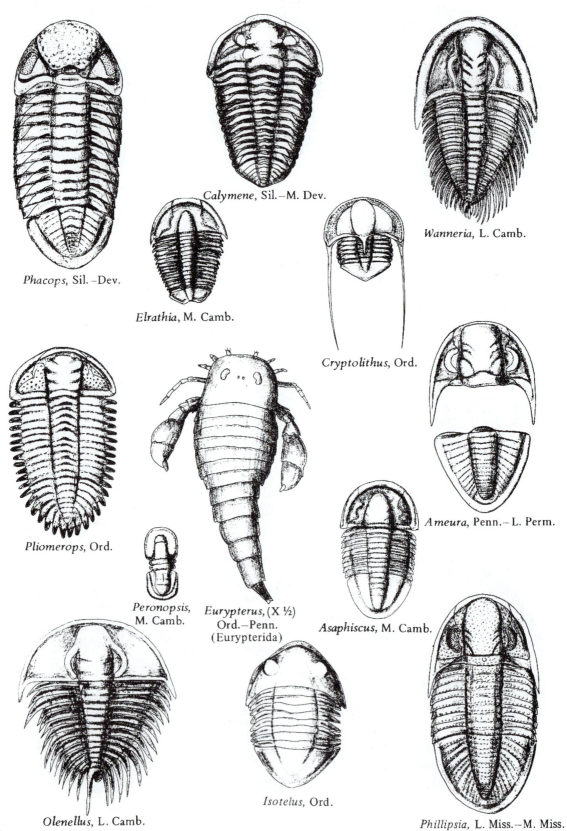

Phacops, Sil.–Dev.

Calymene, Sil.–M. Dev.

Elrathia, M. Camb.

Wanneria, L. Camb.

Cryptolithus, Ord.

Pliomerops, Ord.

Peronopsis, M. Camb.

Eurypterus, (X ½) Ord.–Penn. (Eurypterida)

Asaphiscus, M. Camb.

Ameura, Penn.–L. Perm.

Olenellus, L. Camb.

Isotelus, Ord.

Phillipsia, L. Miss.–M. Miss.

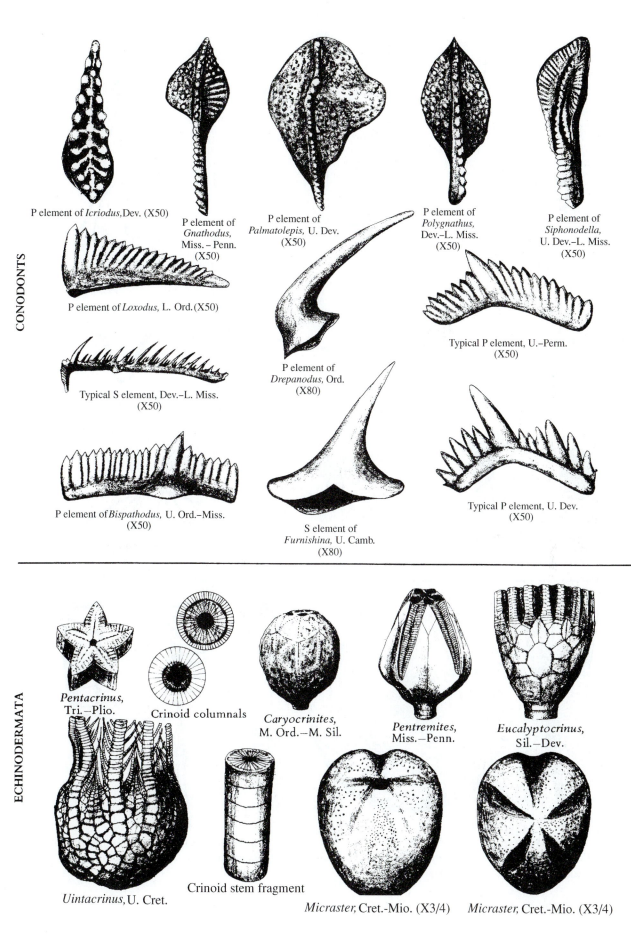

CONODONTS

P element of *Icriodus,* Dev. (X50)

P element of *Gnathodus,* Miss. – Penn. (X50)

P element of *Palmatolepis,* U. Dev. (X50)

P element of *Polygnathus,* Dev.–L. Miss. (X50)

P element of *Siphonodella,* U. Dev.–L. Miss. (X50)

P element of *Loxodus,* L. Ord.(X50)

Typical S element, Dev.–L. Miss. (X50)

P element of *Drepanodus,* Ord. (X80)

Typical P element, U.–Perm. (X50)

P element of *Bispathodus,* U. Ord.–Miss. (X50)

S element of *Furnishina,* U. Camb. (X80)

Typical P element, U. Dev. (X50)

ECHINODERMATA

Pentacrinus, Tri.–Plio.

Crinoid columnals

Caryocrinites, M. Ord.–M. Sil.

Pentremites, Miss.–Penn.

Eucalyptocrinus, Sil.–Dev.

Uintacrinus, U. Cret.

Crinoid stem fragment

Micraster, Cret.-Mio. (X3/4)

Micraster, Cret.-Mio. (X3/4)

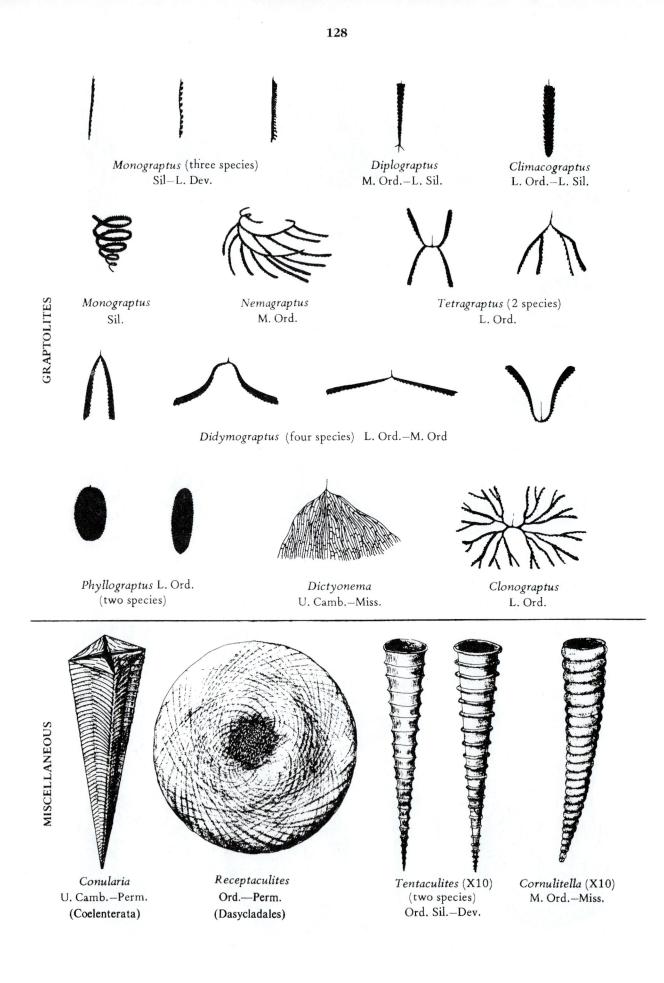

GRAPTOLITES

Monograptus (three species)
Sil–L. Dev.

Diplograptus
M. Ord.–L. Sil.

Climacograptus
L. Ord.–L. Sil.

Monograptus
Sil.

Nemagraptus
M. Ord.

Tetragraptus (2 species)
L. Ord.

Didymograptus (four species) L. Ord.–M. Ord

Phyllograptus L. Ord.
(two species)

Dictyonema
U. Camb.–Miss.

Clonograptus
L. Ord.

MISCELLANEOUS

Conularia
U. Camb.–Perm.
(Coelenterata)

Receptaculites
Ord.–Perm.
(Dasycladales)

Tentaculites (X10)
(two species)
Ord. Sil.–Dev.

Cornulitella (X10)
M. Ord.–Miss.

PLANTS

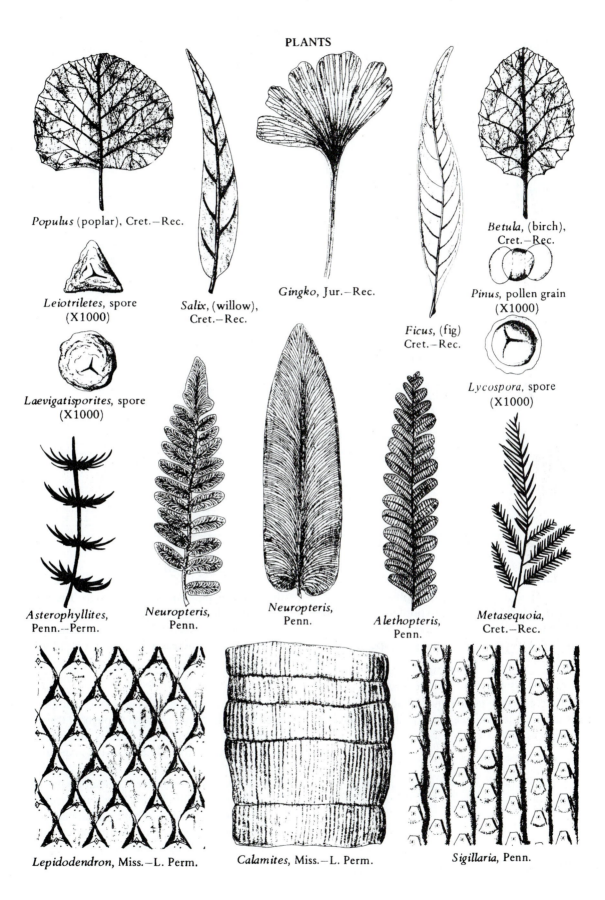

Populus (poplar), Cret.—Rec.

Leiotriletes, spore (X1000)

Laevigatisporites, spore (X1000)

Salix, (willow), Cret.—Rec.

Gingko, Jur.—Rec.

Ficus, (fig) Cret.—Rec.

Betula, (birch), Cret.—Rec.

Pinus, pollen grain (X1000)

Lycospora, spore (X1000)

Asterophyllites, Penn.—Perm.

Neuropteris, Penn.

Neuropteris, Penn.

Alethopteris, Penn.

Metasequoia, Cret.—Rec.

Lepidodendron, Miss.—L. Perm.

Calamites, Miss.—L. Perm.

Sigillaria, Penn.

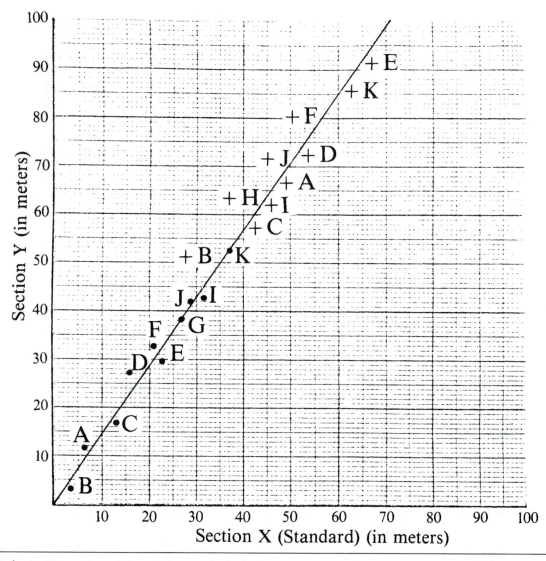

Figure 13.4 An illustration of the graphic method of biostratigraphic correlation using hypothetical data. The starting point for each species, A through K, is plotted with a dot, and the end point for the same species A through K is plotted with a plus sign. The line of correlation is approximated to fit the points. On this chart, the line indicates a time correlation between section X at 37 meters and section Y at 53 meters.

2. When all genera have been identified and the geologic ranges charted, determine the ages for each of the faunal assemblages assigned.

Part C

Alan B. Shaw, an innovative American petroleum geologist, devised a graphic system of biostratigraphic correlation. The procedure is described as follows: using one section of fossiliferous rocks as a standard, carefully record the measured level of the first appearance and the last appearance for each species of fossil throughout the section. Similar data are recorded for all species in common at another stratigraphic section. Data from the standard section are plotted on the horizontal, or *X* axis, and corresponding data from the second section are recorded on the vertical, or *Y* axis. The points will cluster along a line, called the **line of correlation.** Once this line has been established, it is possible to correlate precise points from one section to the other. By accumulating data in the standard section as additional local sections are correlated, a **composite standard** is generated that has great power in correlating new sections as they are studied. Figure 13.4 illustrates the method of graphic correlation.

1. Using stratigraphic data provided in table 13–1a, plot the first and last appearances for each species A through K on the graph sheet provided (fig. 13.5).

2. What stratigraphic level in section X corresponds to level 85 meters in section Y?

3. Repeat the plotting procedure in question 1 using data provided in table 13–1b. What is the explanation for the horizontal part of the line?

4. Repeat the procedure again using data provided in table 13–1c. What is the explanation for the change in the slope of the correlation line?

Table 13–1a–c Data to Plot in Answering Questions 1, 3, and 4 in Part C of Exercise 13.

Table 1a					Table 1b					Table 1c				
	Start		End			Start		End			Start		End	
Species	X	Y	X	Y		X	Y	X	Y		X	Y	X	Y
A	15	5	49	23	A	14	20	42	50	A	5	14	68	71
B	21	14	94	53	B	6	6	47	50	B	4	5	40	50
C	27	12	99	58	C	21	25	36	47	D	15	24	50	58
D	16	10	80	47	D	15	13	80	50	E	25	34	46	53
E	32	18	83	54	E	22	33	85	58	F	26	38	77	73
F	42	28	91	63	F	32	31	90	54	G	33	47	56	59
G	54	24	68	38	G	7	14	75	50	H	32	42	70	68
H	53	34	79	58	H	25	26	82	53	I	37	50	58	64
I	57	32	86	51	I	70	50	90	63	J	45	55	77	77
J	69	34	85	43	J	36	40	81	52					
K	43	23	81	51	K	40	47	86	53					

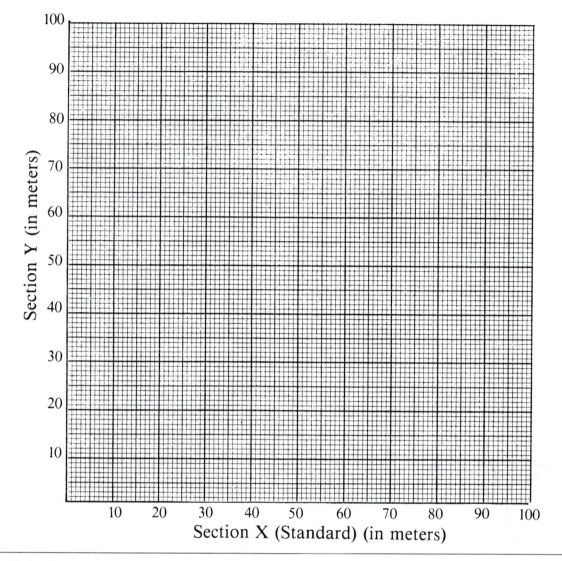

Figure 13.5 Chart to plot the results of questions 1–4 in Part C of exercise. Use different colored pencils or different patterns for the points to separate your answers for questions 1, 3, and 4.

Biostratigraphic Dating Worksheet

Geologic Range / Genera	Ꞓ	O	S	D	M	₱	P	TR	J	K	T	Q

Biostratigraphic Dating Worksheet

Geologic Range											
Genera	€	O	S	D	M	₽	P	₽	J	K	T	Q

135

Biostratigraphic Dating Worksheet

Geologic
Range

Genera

	€	O	S	D	M	₱P	P	₮R	J	K	T	Q

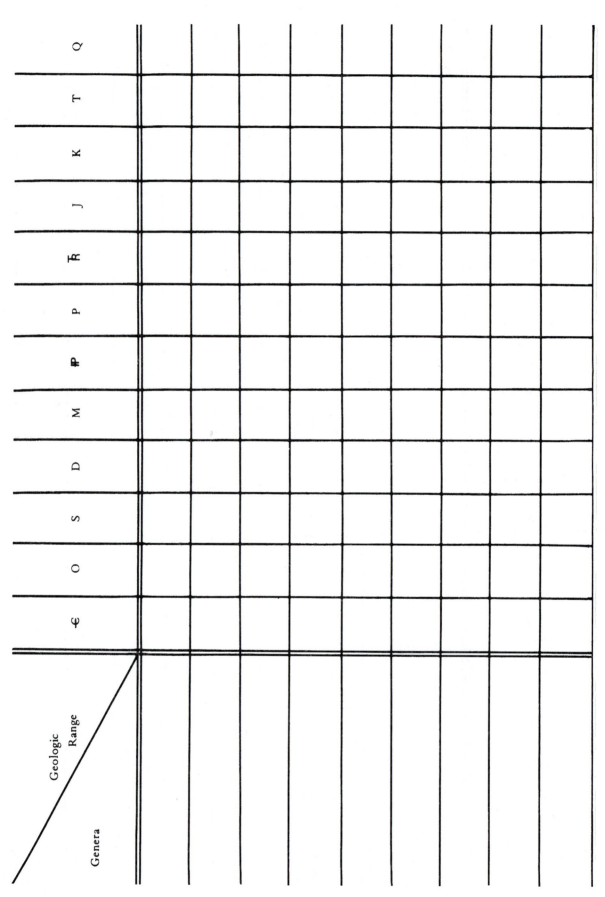

Biostratigraphic Dating Worksheet

Geologic
Range

Genera

€ O S D M P ℞ J K T Q

137

Biostratigraphic Dating Worksheet

Geologic
Range

Genera

€ O S D M ₽ P Ͱ J K T Q

Biostratigraphic Dating Worksheet

Genera	Geologic Range	€	O	S	D	M	₽	P	₮	J	K	T	Q

Biostratigraphic Dating Worksheet

Geologic
Range

Genera

	€	O	S	D	M	₱	P	Ᵽ	J	K	T	Q

Biostratigraphic Dating Worksheet

Geologic Range

Genera

	€	O	S	D	M	₽P	P	Ŧ̶R	J	K	T	Q

The Dinosaurs

From middle Triassic to the end of the Cretaceous, spanning nearly 150 million years, dinosaurs dominated life on land. These animals were the largest to have lived on this planet. The reptilian group called **archosaurs** include alligators and crocodiles as well as their ancestors, pterosaurs, the flying reptiles, the Triassic thecodonts, and the dinosaurs. Dinosaurs are classified into two categories: the **Ornithischia** (bird-hipped) and the **Saurischia** (reptile-hipped) (fig. 14.1).

The ornithischians (fig. 14.2), which were entirely herbivorous, are subdivided into the duck-billed hadrosaurians, the dome-headed pachycephalosaurids, the plated stegosaurids, the armored ankylosaurids, and the horned ceratopsids. The remaining major group, the saurischians, are divided into two groups: the herbivorous sauropods (fig. 14.3) and the carnivorous theropods (fig. 14.4). The sauropods are the giants of the animal kingdom; the theropods are the most voracious.

Vertebrate paleontologists specializing in dinosaurs recognize from 150 to 250 genera found in approximately 150 sites located in North America, South America, Europe, Central Asia, India, China, Australia, and Africa.

Dinosaurs became extinct at the end of the Cretaceous Period, some 65–66 million years ago. Although not the most devastating of extinction events, the Cretaceous–Tertiary event was perhaps the most dramatic. The K/T boundary, as it is called, brought an end to not only dinosaurs, but also other forms, including marine invertebrate ammonoids. The K/T extinction is a worldwide event, explained by either terrestrial causes, such as widespread volcanism, or extraterrestrial causes associated with an asteroid or comet striking the earth. Such a collision would bring about a pervasive alteration of the earth's atmosphere by significantly increasing the amount of suspended particulate material. This dust would reduce the amount of solar radiation able to penetrate to the earth's surface causing a disastrous long-term decline in surface temperatures.

Dinosaur research has received increased attention in recent years, focusing on not only new discoveries of unknown forms, but also their paleobiology and possible relationship to birds. Although most of the museum specimens in the world are based upon specimens found in the late nineteenth century, recent research is uncovering new forms at the rate of almost one per week. For example, a new type of *Deinonychus,* called "Utahraptor" was discovered in southern Utah early in 1992. "Utahraptor" was 20 feet long and weighed over 1 ton. Of the sauropods illustrated in figure 14.3, *Supersaurus* and *Seismosaurus* were all discovered within the past 20 years.

Procedure

After studying figures 14.2–14.5, illustrating some of the known dinosaurs, answer the following questions:

1. List all evolutionary adaptations you can see from the illustrations that benefit the survival of the herbivores shown in figure 14.2.

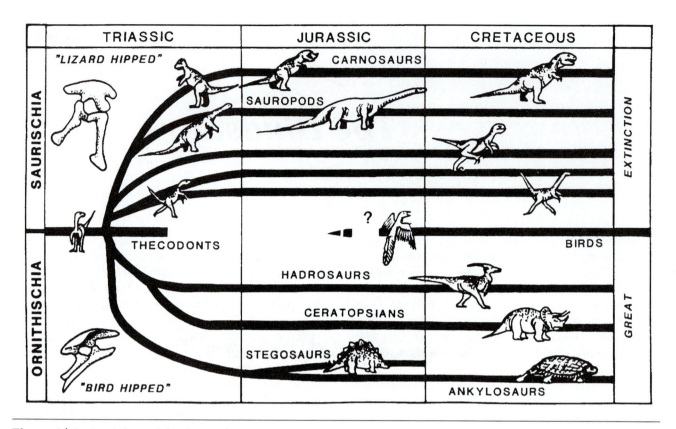

Figure 14.1 Evolution of the dinosaurs. Birds, represented by *Archaeopteryx,* are thought to be possible descendants of dinosaurs. (From Greb, S. F., 1989, Guide to "Progression of Life," Kentucky Geol. Survey, Special Publ. 13, p. 24.)

2. List all evolutionary adaptations you can see from the illustrations that benefit the survival of the sauropods shown in figure 14.3.

3. List all evolutionary adaptations you can see from the illustrations that benefit the survival of the carnivores shown in figure 14.4.

4. What are possible explanations for the unusual head crests of the hadrosaurians shown in figure 14.5?

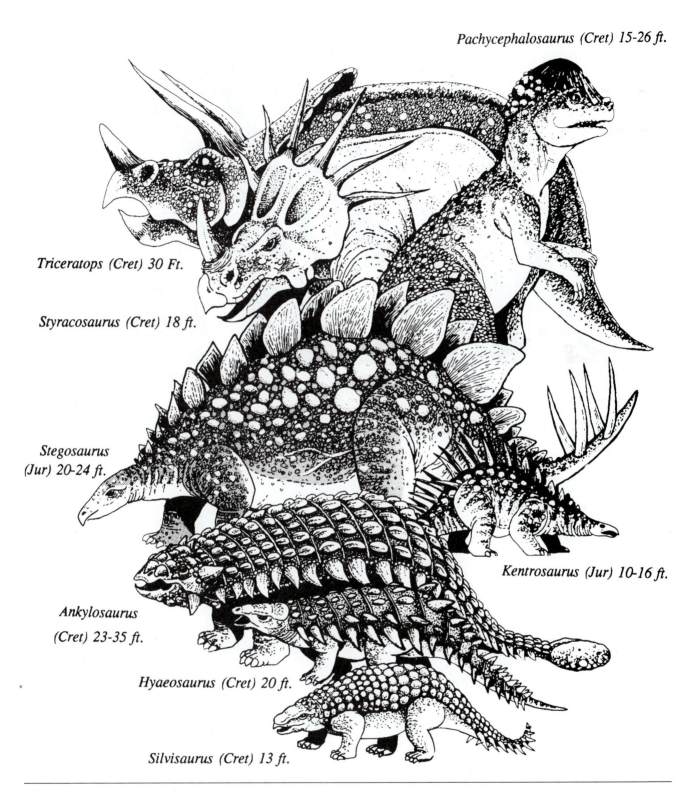

Pachycephalosaurus (Cret) 15-26 ft.

Triceratops (Cret) 30 Ft.

Styracosaurus (Cret) 18 ft.

*Stegosaurus
(Jur) 20-24 ft.*

Kentrosaurus (Jur) 10-16 ft.

*Ankylosaurus
(Cret) 23-35 ft.*

Hyaeosaurus (Cret) 20 ft.

Silvisaurus (Cret) 13 ft.

Figure 14.2 Ornithischian herbivorous dinosaurs from the Jurassic and Cretaceous. (From Greb, S. F., 1989, Guide to "Progression of Life," Kentucky Geol. Survey, Special Publ. 13, p. 28.)

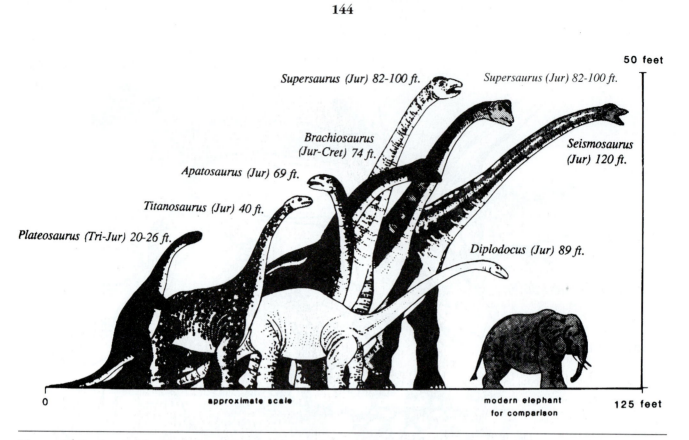

Figure 14.3 Saurischian herbivorous sauropods from the Triassic and Jurassic periods. (From Greb, S. F., 1989, Guide to "Progression of Life," Kentucky Geol. Survey, Special Publ. 13, p. 25.)

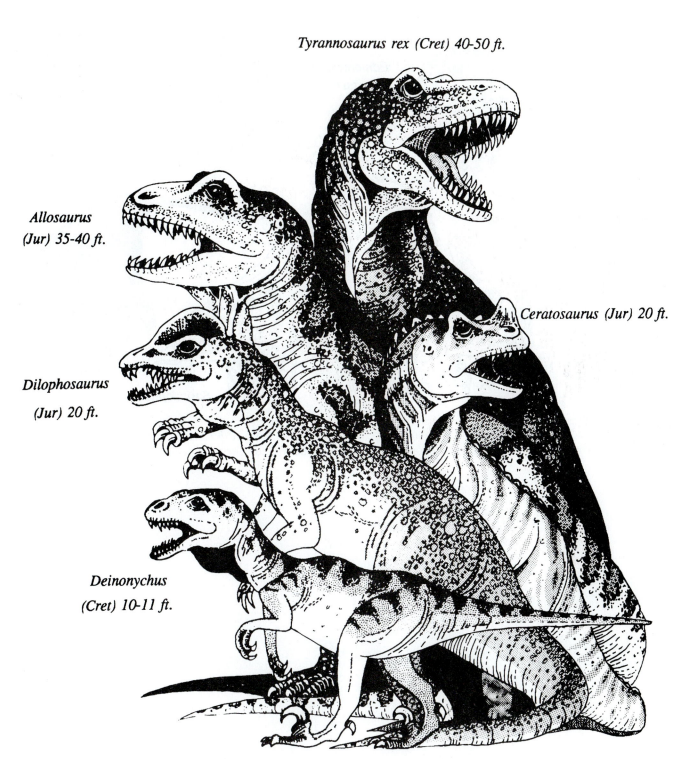

Tyrannosaurus rex (Cret) 40-50 ft.

*Allosaurus
(Jur) 35-40 ft.*

Ceratosaurus (Jur) 20 ft.

*Dilophosaurus
(Jur) 20 ft.*

*Deinonychus
(Cret) 10-11 ft.*

Figure 14.4 Saurischian carnivorous theropods from the Jurassic and Cretaceous. (From Greb, S. F., 1989, Guide to "Progression of Life," Kentucky Geol. Survey, Special Publ. 13, p. 29.)

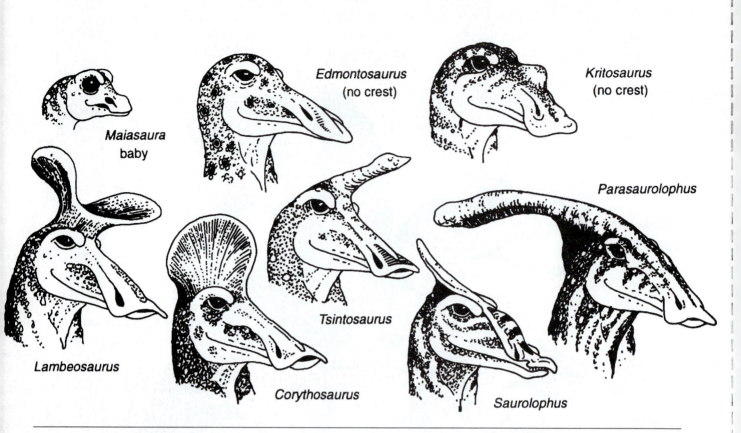

Figure 14.5 Hadrosaurian head crests. All genera illustrated are from the Upper Cretaceous. (From Stephen F. Greb, 1989, Guide to "Progression of Life" Kentucky Geol. Survey, Special Publ. 13, p. 27.)

PART

TWO
Application

Interpretation of Geologic Maps

The following figures provide a set of rules with block diagram illustrations that will aid in the interpretation and understanding of geologic maps. As you examine figures 15.1 thru 15.13, realize that the upper surfaces of the diagrams correspond to the view shown by a geologic map, whereas the sides of the diagrams illustrate the third dimension that the student must visualize from only the map view.

Procedure

Block diagrams are an excellent means of teaching a student to think in three dimensions. This skill is helpful in visualizing the connection between map views and the spacial relationships they represent.

The following block diagrams (figs. 15.14 and 15.15) are for student practice. The first (fig. 15.14), has partially executed diagrams for the student to complete. The second (fig. 15.15), should be completed by the student according to instructions given by the instructor.

On several of the map exercises that follow, the student is instructed to make a geologic cross section that will

allow a sound interpretation of the existing structure. Figure 15.16 is an illustration of the procedures that follow. Study the illustration until you understand how to make geologic cross sections and refer to the procedure as needed to complete future exercises.

Geologic cross sections can be made from geologic maps by the following procedure:

1. Overlay a strip of paper along the specified section line on the geologic map.
2. Mark on the overlay strip the point of intersection of all formation contacts and faults.
3. Determine from the map the direction and angle of dip of each formation and fault, and project the contacts at the dip angle for approximately 5 mm.
4. Connect the projected lines into anticlines or synclines. Interrupt the pattern at the intersection of fault planes.

Common symbols used on geologic maps can be found in figure 15.17.

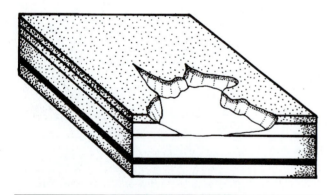

Figure 15.1 Horizontal beds have a broad continuous outcrop pattern that will parallel the contour lines.

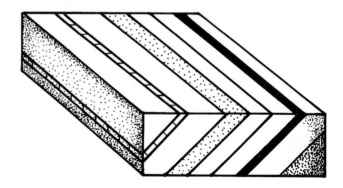

Figure 15.2 Dipping strata will be expressed on a map as stripes, with the youngest strata in the direction of dip in a normal sequence.

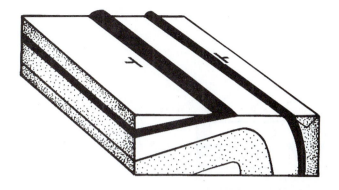

Figure 15.3 For a stratum of constant thickness, the outcrop pattern will broaden with a decrease in dip and shorten with an increase in dip. The strike and dip directions of the beds are shown by the "T-shaped" symbol. The dip angle is measured in degrees.

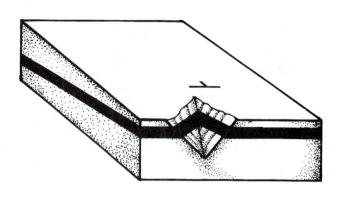

Figure 15.4 As strata intersect stream valleys, the beds will form a "V" pattern with the "V" pointing in the direction of dip. An exception to this occurs when the dip of the bed is less than the slope angle.

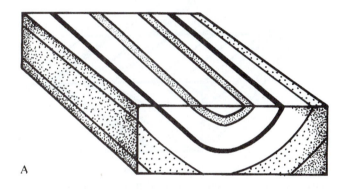

A

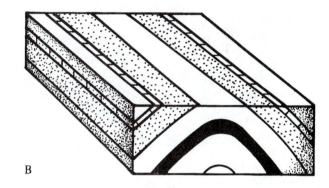

B

Figure 15.5 **A** and **B.** Synclines and anticlines whose axes are horizontal are called horizontal folds.

Horizontal folds will have a parallel outcrop pattern on a map.

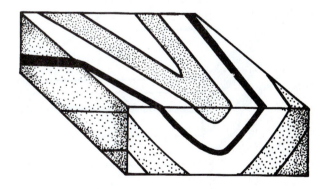

Figure 15.6 Folds whose axes are inclined to the horizontal are called plunging. Plunging folds have converging or diverging outcrop patterns. The outcrop pattern of a plunging anticline will close, or "V," in the direction of plunge.

Figure 15.7 The outcrop patterns of plunging synclines open in the direction of plunge. The youngest beds will be exposed in the center of a syncline. The oldest beds are exposed in the center of an anticline.

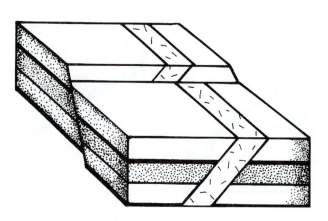

Figure 15.8 Abrupt breaks in outcrop patterns are surface expressions of faults.

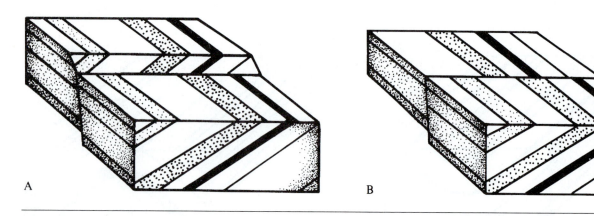

A

B

Figure 15.9 **A** and **B.** When dipping beds are cut by faults, there is an apparent lateral shift in the outcrop pattern toward the direction of dip on the upthrown block.

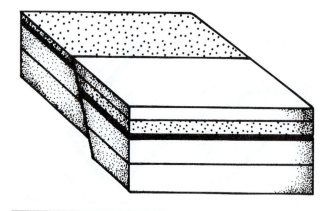

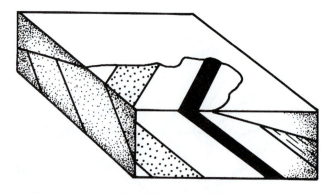

Figure 15.10 Across any fault trace, the oldest beds will be exposed on the upthrown block. Apply this rule to figure 15.9B also.

Figure 15.11 The outcrop pattern of nearly horizontal beds superimposed on the striped outcrop pattern of dipping beds indicates an **angular unconformity.**

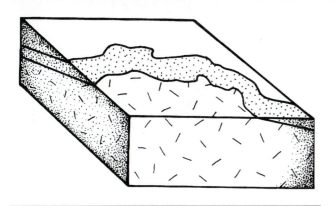

Figure 15.12 Unaltered sedimentary formations directly overlying an intrusive igneous rock mass indicates a **nonconformity.**

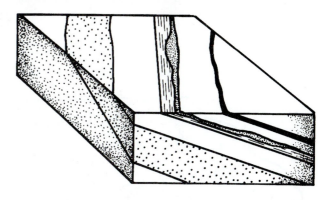

Figure 15.13 The irregular outcrop pattern indicating an ancient erosion surface enclosed by parallel formations with rocks of intervening ages missing illustrates a **disconformity.**

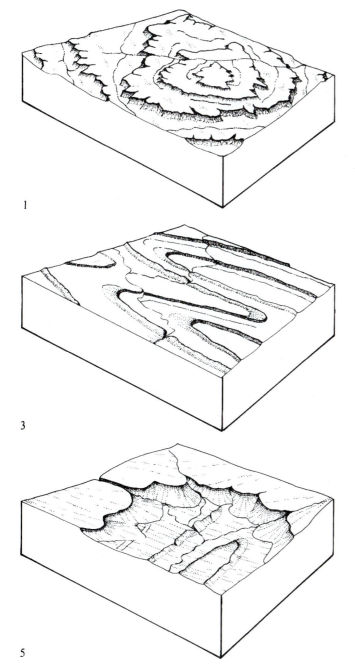

1

2

3

4

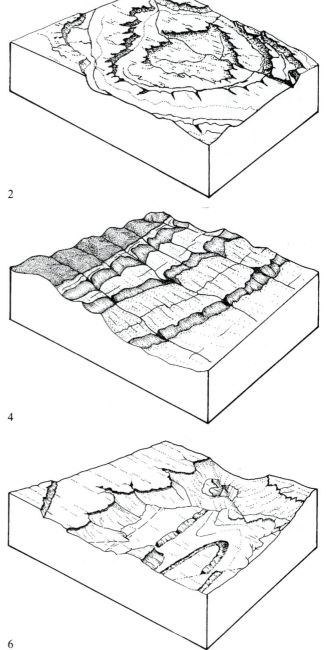

5

6

Figure 15.14 Practice block diagrams.

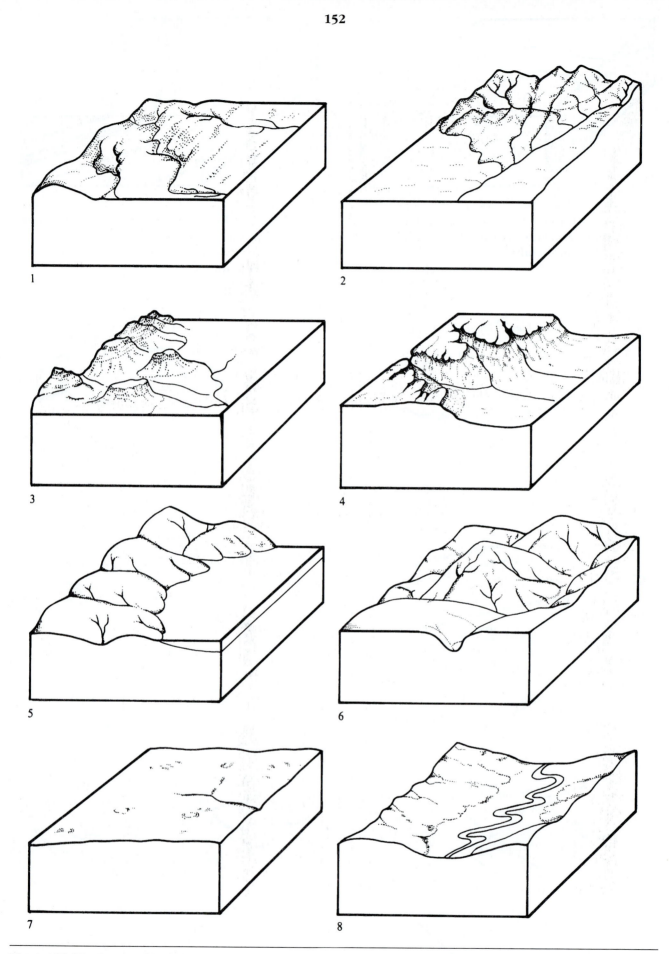

Figure 15.15 Practice block diagrams.

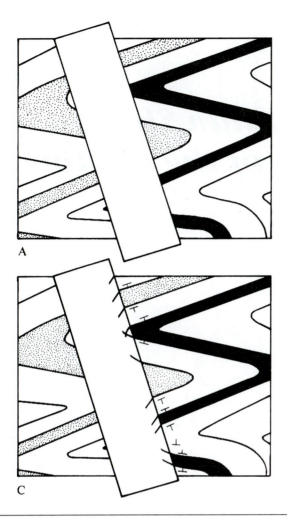

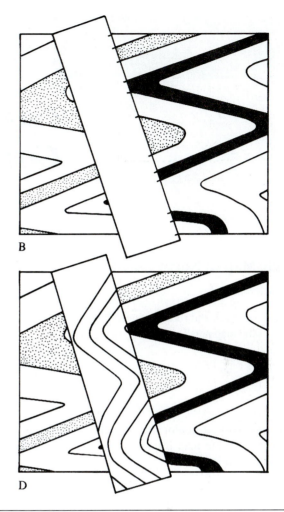

Figure 15.16 Diagrams showing the procedure to be followed in construction of a geologic cross section from map data.

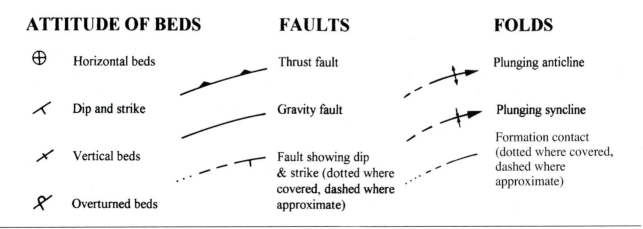

ATTITUDE OF BEDS

⊕   Horizontal beds

⤝   Dip and strike

⤟   Vertical beds

⤪   Overturned beds

FAULTS

Thrust fault

Gravity fault

Fault showing dip & strike (dotted where covered, **dashed where approximate**)

FOLDS

Plunging anticline

Plunging syncline

Formation contact (dotted where covered, dashed where approximate)

Figure 15.17 Common symbols used on geologic maps.

Canadian Shield—
Core of the Continent

Geologic Map of North America

Scale: 1 inch = 80 miles

1. What are the ages and rock types on the north and south halves of the map area?

2. Draw an east-west geologic cross section across the central part of the map from St. Paul, Minnesota, to Hamilton, Ontario.

3. Explain the isolated Cretaceous occurrences south of James Bay in the northeastern corner of the map.

4. What are the geologic factors controlling the position and shape of Georgian and Green bays?

5. Give reason(s) why the map pattern for Silurian rocks is discontinuous.

6. How do the age distribution and map pattern indicate this is a basin?

CANADIAN SHIELD

K	Light green	Cretaceous
P_z	Medium blue gray	Paleozoic
IP_2	Blue and white, diagonal	Upper Pennsylvanian
IP_1	Blue	Lower Pennsylvanian
M_1	Lavender and red, diagonal	Mississippian
D	Brownish gray	Devonian
D_2	Brown	Upper Devonian
D_1	Brownish gray	Lower Devonian
S	Light purple	Silurian
O	Reddish purple	Ordovician
ϵ	Light orange-brown	Cambrian
$p\epsilon u$	Yellowish tan	Upper Precambrian
$p\epsilon u_2$	Reddish brown	Keweenawan sedimentary rocks
$p\epsilon u_1$	Light brown	Keweenawan volcanic rocks
$p\epsilon b$	Brownish green	Precambrian basic intrusives
$p\epsilon m$	Light greenish brown	Middle Precambrian
$p\epsilon i$	Pink	Precambrian granite and granite gneiss
$p\epsilon l$	Olive green	Lower Precambrian

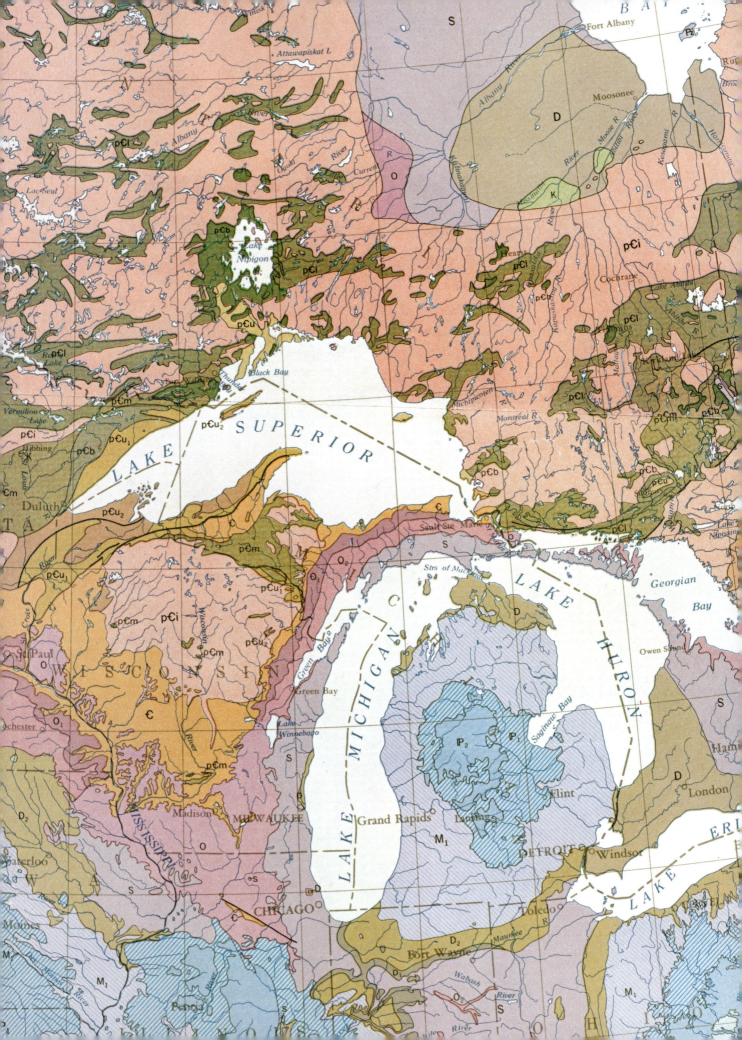

Interior Lowland—
The Stable Interior

Geologic Map of North America
(See exercise 16)

Scale: 1 inch = 80 miles

1. What is the general structure of the rocks in Michigan and Wisconsin?

2. Why are Ordovician, Silurian, and Devonian rocks missing in the area east of Lake Superior?

Geologic Map of the United States
(See exercise 21)

Scale: 1 inch = 40 miles

3. Draw a geologic cross section from the northwestern corner of Georgia to the northwestern corner of the map.

4. What is the geologic structure of central Tennessee near Nashville?

EXERCISE 18

Folded Appalachian Mountains

Geologic Map of Pennsylvania

Scale: 1 inch = 4 miles

1. What is the likely source area for the Ordovician Martinsburg Formation and the Devonian Catskill Formation? With what geologic events are these associated?

2. Account for the difference in the structural patterns north and south of the Pennsylvania Turnpike, which extends east-west through Carlisle, Pennsylvania.

3. Date the Appalachian folding in this area. Summarize the evidence.

4. Explain the origin and distribution of the igneous rocks in this area. With what episode of diastrophism are they associated?

5. What is the reason for the present course of the Susquehanna River cutting across the folds indiscriminately and across all different rock types?

6. In terms of plate tectonics, what type of plate interaction best explains the geology of the map area?

APPALACHIANS

TRIASSIC

Ŧ d	Red diamonds and red lines	Dark gray igneous sills and dikes
Ŧ lc		
Ŧ g	Bright green with red patterns	Brunswick or Gettysburg Formation
Ŧ b		Red sandstone and shale with minor conglomerate
Ŧ h		
Ŧ ac		

PENNSYLVANIAN

IPp	Blue green	Pottsville Group
		Sandstone, conglomerate, shale and coals

MISSISSIPPIAN

Mmc	Pink	Mauch Chunk Formation
		Gray shale and red sandstone
Mp	Blue purple	Pocono Group
		Conglomerate, sandstone, and shale

DEVONIAN

Dck	Orange	Catskill Formation
		Red sandstone and shale
Dm	Light orange	Marine beds, shale, sandstone, and limestone
Dho	Gray, green diagonal	Hamilton Group and Onendaga Formation, Shale, sandstone, and limestone
Doh	Orange and red diagnoal	Oriskany and Helderberg Formations
		Fossiliferous sandstone and shale, and fossiliferous limestone

SILURIAN

Skt	Purple and blue diagonal	Keyser Formation and Tonoloway Formations
		Limestone
Sw	Dark and light blue, horizontal	Wills Creek Fromation
		Shale and limestone with local sandstone
Sc	Light green, red stipples	Clinton Group
		Red iron-rich shale and sandstone
Sbm	Blue and purple, diagonal	Bloomsburg and McKenzie Formation
		Red to green shale and sandstone
St	Red and brown, diagonal	Tuscarora Formation
		Light coarse sandstone

ORDOVICIAN

Ojb	Orange and brown, diagonal	Juniata Formation and Bald Eagle Formation
		Red sandstone, shale, and conglomerate
Om	Gray and pink, horizontal	Martinsburg Formation
		Marine shale
Oc		
Ohm	Blue grayish pink, diagonal	Chambersburg Formation, Hershey and Myerstown Formations
Osp	Light blue	St. Paul Group and Annville Formation
		Limestone with chert
Ob	Coarse pink and purple, diagonal	Beekmantown Group
		Limestone
Oor	Fine pink and purple, diagonal	Ontelaunee Formation, Epler Formation
		Rickenback Formation
		Limestone and dolomite
Os	Pink with blue, diagonal	Stonehenge Formation
		Limestone and limestone conglomerate

CAMBRIAN

€c	Red and bright green, diagonal	Conococheague Group
		Limestone and dolomite
€e	Grayish green, orange, crossed lines	Elbrook Formation
		Limestone and dolomite
€wb	Green and orange vertical	Waynesboro Formation
		Red and purple shale with sandstone beds
€t	Red and light green, diagonal	Tomstown Formation
		Dolomite, thin shaley beds
€a	Pink and red squares	Antietam Formation
		Quartzite and schist
€ma	White and orange diagonal	
€h	Orange and red vertical	Harpers Formation
		Phyllite and schist
€wl	Pink and blue diagonal	Chickies Formation or Weverton Formation
		Quartzite and schist

IGNEOUS ROCKS

mr	Yellow, red dots	Metarhyolite
vs	Green, red stippled	Greenstone schist
mb	Brown, red stippled	Metabasalt

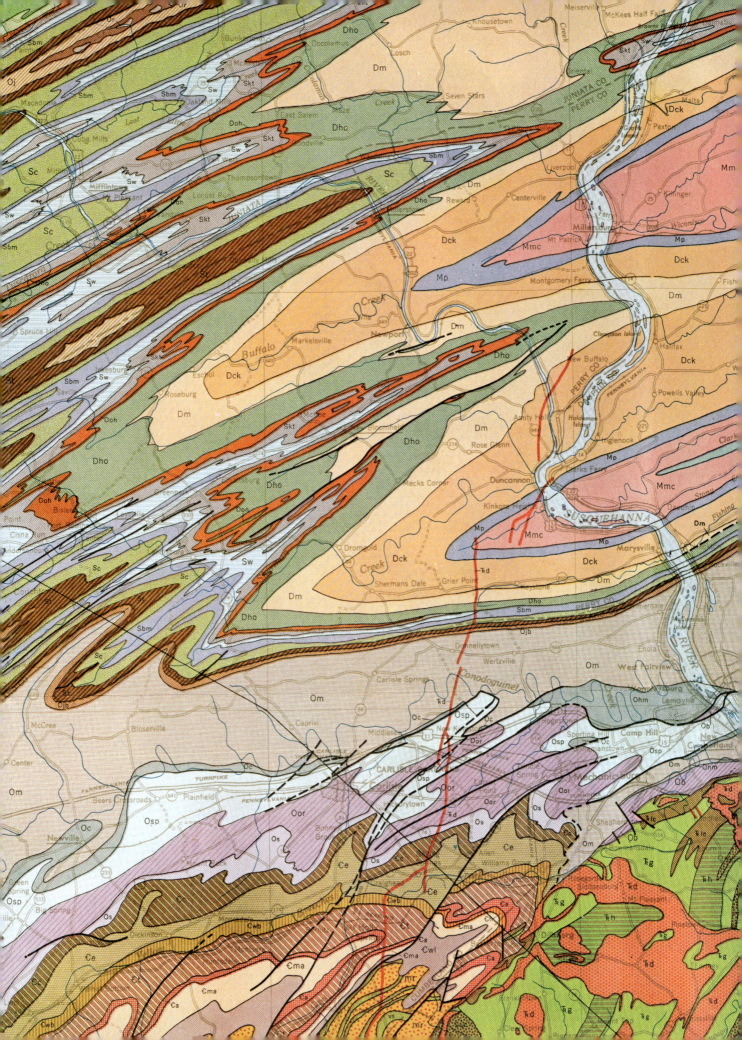

New England Appalachians

Geology of Danforth Quad. Maine

Scale: 1 inch = 1 mile

1. What are the four principal kinds of rocks exposed in the map area? What ages of rocks are exposed in this area?

2. Draw a geologic cross section along the diagonal line from the southern to the northwestern corners of the area.

3. Determine a sequence of geologic events for this area.

4. What episode of orogenic activity in the history of the Appalachian mobile belt is represented on this map?

5. Were these rocks the products of a convergent, divergent, or transform margin? What evidence best supports your answer?

EXERCISE 19

BEDROCK GEOLOGY OF THE DANFORTH QUADRANGLE, MAINE

DEVONIAN		
db	Red	Diabase dike
Dg	Pink	Granite and quartz granodiorite
SILURIAN		
Ss	Light reddish purple	Slate and siltstone
Sq	Dark reddish purple	Quartzite, conglomerate, slate, and siltstone
Sss	Light purple	Slate and siltstone
Sl	Bright purple	Limestone conglomerate
ORDOVICIAN		
Os	Bluish purple	Slate, shale, siltstone, and tuff
Oss	Blue	Slate and siltstone

The Catskills

Geologic Map of New York State

Scale: 1 inch = 4 miles

1. What is the probable source area for the broad expanse of Devonian rocks in the western part of the map area? Compare with figure 5.5.

2. What is the separate evidence for the Taconic and Acadian orogenies in this area, particularly near Becraft Mountain, east of Catskill?

3. Silurian rocks crop out in the northwestern corner of the map area, but are absent elsewhere. Explain this distribution.

5. What type of plate margin interaction would produce the geologic structures seen on this map?

4. What geologic factors control the position of the Hudson River?

169

THE CATSKILLS

QUATERNARY

Q	Lavender	Glacial and alluvial deposits

DEVONIAN

Djwl	Light green and red diagonal	Upper Katsberg Formation
		Red shale, sandstone, and conglomerate
Dsd	Light blue and gray, red stippled	Lower Katsberg Formation
		Red sandstone, shale, and siltstone
Dss	Light blue diagonal with red stipple	Stoney Clove Formation
		Sandstone
Dgk	Green with red stipple	Oneonta and Kaaterskill Formations
		Red shale and sandstone
Dha	Light yellow green, red stipple	Kiskatom Formation
		Shale and sandstone
Dhp	Light yellow green with olive, vertical	Panther Mountain Formation
		Shale, siltstone, and sandstone
Dhm	Light yellow green with olive diagonal	Lower Hamilton
		Shale and siltstone
Dou	Light blue gray squares	Onondaga Limestone, Schoharie Formation
		Limestone and shale
Do	Blue green with red stipple	Oriskany Formation
		Sandstone
Dhg	Gray and blue squares	Helderberg Group
		Limestone and dolomite
DS	Purple and gray diamonds	Silurian and Lower Devonian undifferentiated
		Limestone and shale

SILURIAN

Sbs	Gray with red diagonals	Cobleskill Limestone and undifferentiated
		Bertie and Salina Groups
		Limestone and shale

ORDOVICIAN

Osc	Yellow and olive diagonal	Schenectady Formation
		Sandstone and shale
Osh	Yellow and gray, horizontal	Snake Hill Shale
		Black shale
On	Yellow	Normanskill Formation
		Black shale with sandstone
O€	Orange	Lower Ordovician and Upper Cambrian
		Limestone, sandstone, and shale

CAMBRIAN

€c	Olive green	Cossayuna Group
		Black shale

CAMBRIAN(?)

Xe	Wavy olive green	Elizaville Group
		Slate and quartzite

From the Geological Map of New York State, 1961, published by the Geological Survey, New York State Museum and Science Service, Albany.

Coastal Plains

Geologic Map of the United States

Scale: 1 inch = 40 miles

1. What four main geologic provinces are represented in the map area?

2. What ages of rock characterize each province?

3. What is the general structure of each province?

4. Describe the nature of the contact between the folded Appalachians and the Coastal Plains.

5. Draw a geologic cross section from 1 cm north of the "B" in Alabama southwest to the southwestern corner of the map.

6. Explain the difference between the formation of the Appalachian Mountains and the Coastal Plains.

COASTAL PLAIN

QUATERNARY

| Qs | Light blue | Coastal and estuarine sand and gravel |

PLIOCENE

| Pc | Tan with orange circles | Fluvial sediments |

MIOCENE

| Mt | Light orange | Tampa and Catahoula Formations Marl and sand sediments |

OLIGOCENE

| Φv | Dark and light yellow diagonal | Vicksburg Group |

EOCENE

Ej	Light yellow	Jackson Group Marl
Ec	Dark and light yellow, horizontal	Claiborne Group
Ew	Fine light and dark yellow, horizontal	Wilcox Group
Em	Yellow-orange	Midway Group

CRETACEOUS

Kr	Fine green vertical	Ripley Formation
Ks	Coarse green vertical	Selma Chalk
Ke	Coarse green horizontal	Eutaw Formation
Kt	Coarse green diagonal	Tuscaloosa Formation

PENNSYLVANIAN

| Cpv | Light gray | Pottsville Group |

MISSISSIPPIAN

Cm	Dark blue gray	Mississippian Undivided
Cmu	Light blue gray diagonal	Upper Mississippian Undivided
Cmm	Medium blue stipple	Middle Mississippian undifferentiated
Cml	Medium blue horizontal, with stipple	Lower Mississippian undifferentiated
Cg	Purple triangles	Middle Paleozoic granite

DEVONIAN

| D | Medium purple | Devonian |

SILURIAN

| S | Light lavender | Silurian |
| SO | Purple, horizontal | Silurian-Ordovician |

ORDOVICIAN

| Ou | Light lavender | Upper Ordovician |
| Om | Fine purplish red, horizontal | Middle Ordovician |

CAMBRIAN

€O	Light orange	Cambrian-Ordovician
€l	Dark orange, brown diagonal	Lower Cambrian
€q	Orange and brown stipple	Lower Cambrian Quartzites and slates

METAMORPHOSED PRECAMBRIAN AND PALEOZOIC

Awh	Brown, fine white dashes	Wissahickon Schist, High Rank
Awl	Irregular brown diagonal	Wissahickon Schist, Low Rank
Awi	Brown, red overlay	Wissahickon Schist with igneous masses
Agr	Orange	Intrusive rocks and gneiss
Agg	Irregular orange diagonal	Granite gneiss
Acs	Brown	Metamorphosed sedimentary and volcanic rocks

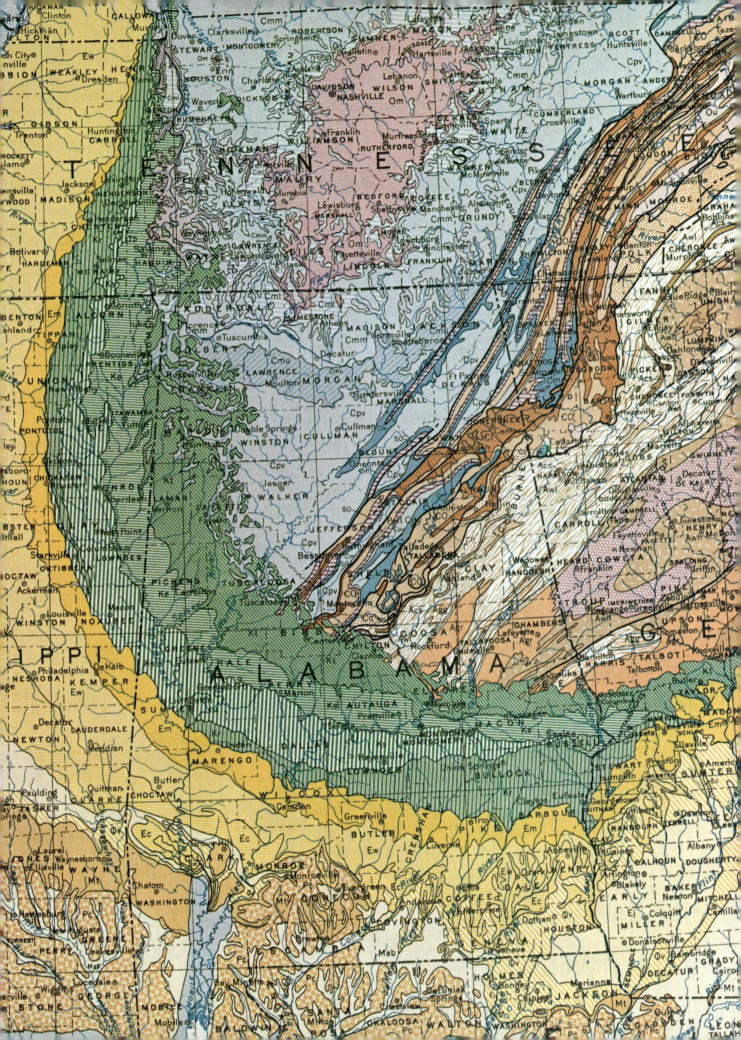

Colorado Plateau and Ancient Colorado Mountains

Geology of the Gateway, Colorado Quad.

Scale: 1 inch = 2000 feet (approximately 0.4 miles)

1. Draw a geologic cross section along the diagonal line from the northeastern corner to the south-central part of the map.

2. Note the unconformable contact between the Cutler Formation and the Precambrian gneiss and schist. The Cutler Formation is missing in the northeastern corner. Where exposed on the map, the Cutler Formation is conglomeratic. In light of these observations, explain the origin of the Cutler Formation.

3. Explain the absence of the Moenkopi Formation beneath the Chinle Formation in the northeastern corner of the map.

4. What is the structure of the Mesozoic rock units in the southern quarter of the map area?

5. Determine a sequence of events for the rock units in the map area. What is the age of the fault along Birch Creek?

6. What is the difference between the style of deformation shown here and that shown in the Canadian Rocky Mountains (exercise 25)?

GATEWAY QUADRANGLE, COLORADO

QUATERNARY

Qal	Stippled yellow	Alluvium
Qg	Green, yellow circles	Terrace gravels
Qfg	Dark yellow stipple	Angular fragments and boulders

CRETACEOUS

Kbc	Dark blue green diagonal	Burrow Canyon Formation
		Sandstone and conglomerate with interbedded green and purplish shale

JURASSIC

Jmb	Green horizontal	Morrison Formation
Jms	Solid blue green	Upper member of rusty red to gray shale
		Lower member of varicolored shale with buff sandstone and lenses of conglomerate
Js	Olive green, vertical	Summerville Formation
		Gray, green, and brown sandy shale and mudstone
Jec	Dark olive green, diagonal	Entrada Sandstone and Carmel Formation undivided
		Fine-grained massive sandstone and red mudstone
Jn	Dark olive green, solid	Navajo Sandstone
		Buff, gray, cross-bedded sandstone
Jk	Olive green, horizontal	Kayenta Formation
		Interbedded red, gray siltstone and sandstone
Jw	Dark olive, diagonal	Wingate Sandstone
		Massive reddish brown cross-bedded sandstone

TRIASSIC

$\bar{R}$ c	Light blue, diagonal	Chinle Formation
		Red siltstone with lenses of red sandstone, shale, and conglomerate
$\bar{R}$ ml	Dark blue, horizontal	Moenkopi Formation
		Sandy mudstone, brown and red with local gypsum

PENNSYLVANIAN

Pc	Medium blue, horizontal	Cutler Formation
		Red to purple conglomerate and sandstone

PRECAMBRIAN

p€	Brown and white, mottled	Gneiss, schist, granite, and pegmatite

Wasatch Mountains

Geology of Brighton and Dromedary Peak Quads., Utah

Scale: Approximately 2 1/2 inches = 1 mile

1. Study the ages of rocks that have been deformed by faults or folds. What is the earliest possible time for the tectonic event responsible for these features?

2. What is the name of the orogenic event whose effects are shown on this map?

3. What type of unconformity exists between the Fitchville Formation and the Maxfield Limestone?

4. Are the tectonic features the result of compression, tension, or shear?

5. Locate the old historic mining town of Alta, Utah. What is the likely source for the silver ore mined at Alta in the late 1800s? When were the silver-bearing minerals deposited in the rock?

WASATCH MOUNTAINS

QUATERNARY

Qal	Light yellow	Stream gravel and valley fill
Qt	Light yellow, black dots	Talus
Qls	Yellow	Landslide debris
Qm	Light yellow, pink circles	Glacial deposits

TERTIARY

pd	Bright purple	Pebble dikes
ld	Bright green	Lamprophyric dikes
sd	Bright orange	Silicic dikes
md	Bright red	Intermediate dikes
cp	Pink	Granodiorite of Clayton Peak Stock
ap	Dark pink, white dashes	Porphyritic granodiorite of Alta Stock
ag	Dark pink	Granodiorite of Alta Stock

PENNSYLVANIAN

IPrv	Light blue gray	Round Valley Limestone
		Gray limestone with chert and fossils

MISSISSIPPIAN

Mdo	Dark gray blue	Doughnut Formation
		Thin-bedded, dark gray, fine-grained limestone with black shale
Mh	Light blue	Humbug Formation
		Limestone and sandstone
Md	Medium bluish purple	Deseret Limestone
		Limestone and dolomite with chert
Mdg	Medium bluish pink	Deseret and Gardison Limestone
Mg	Medium gray blue	Gardison Limestone
		Fossiliferous limestone
Mf	Bright blue	Fitchville Formation
		Dark gray to white dolomite

CAMBRIAN

€m	Light brown	Maxfield Limestone
		Dark to light gray mottled
		Limestone, some dolomite and shale
		€ md identifies a thin (5–15 foot)
		bed of white dolomite
		at top of Maxfield Ls.
€o	Medium orange brown	Ophir Formation
		Olive micaceous shale with limestone
€t	Light orange brown	Tintic Quartzite
		Light brown orthoquartzite
p €mf	Gray	Mineral Fork Tillite
		Pebbles and boulders in sandstone, some shale
p €bc	Medium gray	Big Cottonwood Formation
		Thick beds of quartzite and shale

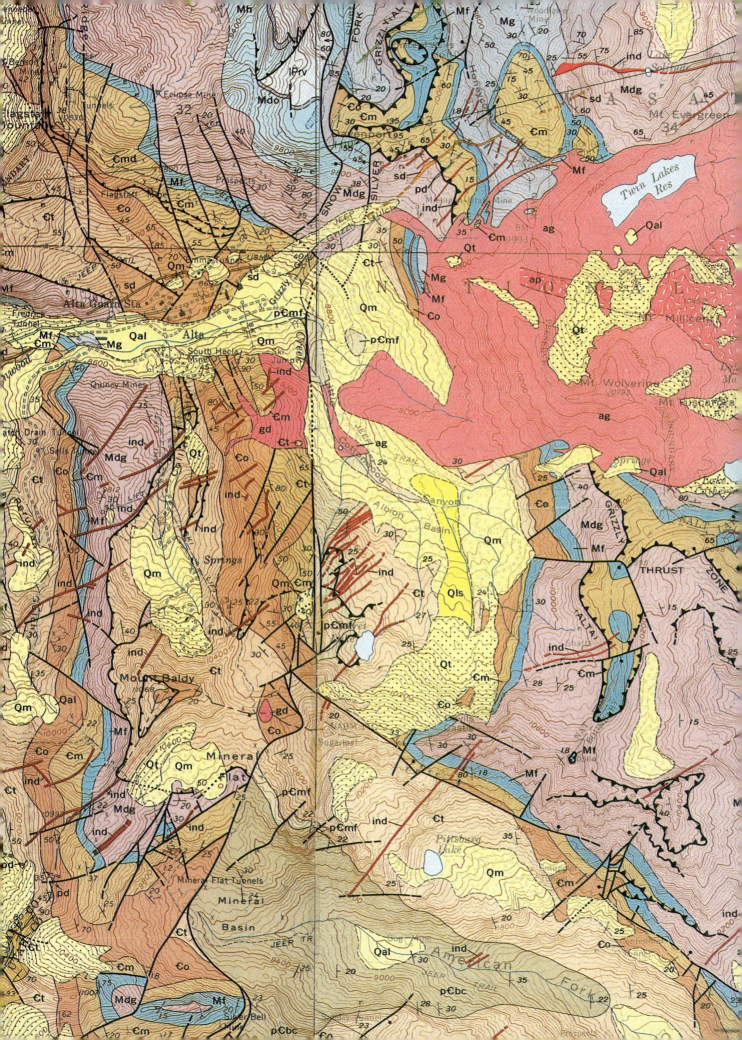

Uinta Mountains

Geologic Map of Utah

Scale: 1 inch = 4 miles

1. What is the general structure and trend of the Uinta Mountains?

2. What is the structural significance of the large block of Pennsylvanian-Permian Oquirrh Formation posed in the southwestern corner of the map?

3. What is the age of the large fault in the south-central part of the map area approximately 1 mile southwest of Current Peak? (The dashed symbol on this fault shows the approximate position of the buried fault.)

4. What are the ages of faulting in the northwestern corner of the map area?

5. What is the probable source for the coarse conglomerates along the northern edge of the map?

6. Where is the most likely area for metallic mineral deposits? Why?

8. What is the name for the tectonic event that developed these mountains?

7. What is the probable origin of the large exposure of Tertiary volcanics in the central part of the map?

9. Is the tectonism present reflected primarily as compressional features or as vertical uplift?

UINTA MOUNTAINS

QUATERNARY

Qay	Gray	Recent young alluvium
Qao	Gray stippled	Recent old alluvium
Qls	Yellow with irregular gray lines	Landslide
Qgm	Yellow with gray circles	Moraine
Ogo	Yellow	Glacial outwash

TERTIARY

Tgs	Yellow orange	Gravel terraces
Tig	Pink with red dashes	Granitic intrusive masses
Tib	Red	Basic intrusive masses
T₁ap	Light pink with gray circles	Intermediate volcanic breccias and flows

NORTH FLANK OF UINTAS

TERTIARY

Tk	Orange with pink circles	Knight Conglomerate
		Conglomerate with sandstone

CRETACEOUS

Kec	Light green with red circles	Echo Canyon Conglomerate
		Sand, shale, and conglomerate
Kw	Light and medium green diagonals	Wanship Formation
		Marine shale and sandstone
Kf	Light green and dark green stippled	Frontier Sandstone
		Sand with coal
Ka	Dark green diagonal	Aspen Shale
		Gray marine shale
Kk	Light green with dark green coarse stipple	Kelvin Conglomerate
		Red fluvial sediments

SOUTH FLANK OF UINTAS

TERTIARY

Tdr	Pinkish orange diagonal	Duschene River Formation
		Fluvial sediments
Tu	Light pinkish orange	Uinta Formation
		Fluvial sediments

CRETACEOUS

Kcc	Green with pink circles	Current Creek Formation
		Fluvial sediments
Kmv	Light yellow green	Mesa Verde Formation
		Sand, shale, and coal
Kms	Light blue green	Mancos Shale
		Dark marine shale
Kmf	Light yellow and green diagonals	Mowry and Frontier Formations
		Shale, sand, and coal
Kdcm	Green and gray, vertical	Dakota and Cedar Mountain Formations
		Sand and shale

ENTIRE MAP

JURASSIC

Ju	Green diamonds	Undiff. Jurassic
Jm	Medium blue and red diagonal	Morrison Formation
		Red, gray, green fluvial sediments with dinosaur remains
Jst	Light blue green and red diagonal	Stump Sandstone
		Green and brown sandstone
Jp	Light blue green and red stipple	Preuss Sandstone
		Red siltstone
Jtc	Light blue green and red lines	Twin Creek Limestone
		Light gray limestone and shale
Jn	Light green with olive stipple	Navajo Sandstone
		Cross-bedded sandstone

TRIASSIC

ᵀℝc	Blue	Chinle Formation
		Mixed fluvial red and green shale and sandstone
ᵀℝs	Blue with olive stipple	Shinarump Conglomerate
		Fluvial conglomerate
ᵀℝa	Light green and dark green diagonal	Ankareh Formation
		Red siltstone and shale
ᵀℝt	Medium blue green and olive, diagonal	Thaynes Formation
		Red siltstone and limestone
ᵀℝw	Blue green diagonal	Woodside Shale
		Red siltstone and shale

PERMIAN

ℙpc	Blue and pink diagonal	Park City Formation
		Limestone with phosphatic shale

SOUTHERN WASATCH MOUNTAINS

PERMIAN

ℙdc	Light blue with blue stipple	Diamond Creek Sandstone
		Light cross-bedded sandstone
ℙk	Light and medium blue, horizontal	Kirkman Limestone
		Thin-bedded limestone

PENNSYLVANIAN

ℙℙo	Light blue, fine stipple	Oquirrh Formation
		Quartzite and limestone
PMmc	Medium blue	Manning Canyon Formation
		Marine shale and limestone

CENTRAL WASATCH MOUNTAIN SECTION

PENNSYLVANIAN

Pw	Blue and coarse pink stipple	Weber Quartzite
		Cross-bedded sandstone
Prv	Blue	Round Valley Limestone
		Limestone

MISSISSIPPIAN

Mdo	Light blue with blue stipple	Doughnut Formation
		Dark limestone and shale
Mn	Pink with blue stipple	Humbug Formation
		Limestone
Md	Coarse blue and pink, diagonal	Deseret Formation
		Limestone and dolomite

UINTA MOUNTAIN SECTION

PENNSYLVANIAN

Pw	Blue, red stipple	Weber Formation
		Cross-bedded sandstone
Pm	Light blue and pink, horizontal	Morgan Formation
		Red sandstone and shale
PMmc	Medium blue	Manning Canyon Formation
		Shale

MISSISSIPPIAN

Mu	Light purple	Upper Mississippian
		Limestone and shale
Ml	Blue and pink diagonal	Lower Mississippian undifferentiated

CAMBRIAN

€t	Gray, coarse red stipple	Tintic Quartzite
		Quartzite

PRECAMBRIAN

p€rp	Dark brown red stipple	Red Pine Shale
		Brown, gray shale
p€m	Brown with red dashes	Mutual Formation
		Purple quartzite
p€lu	Light gray with fine red stipple	Uinta Group
		Quartzite

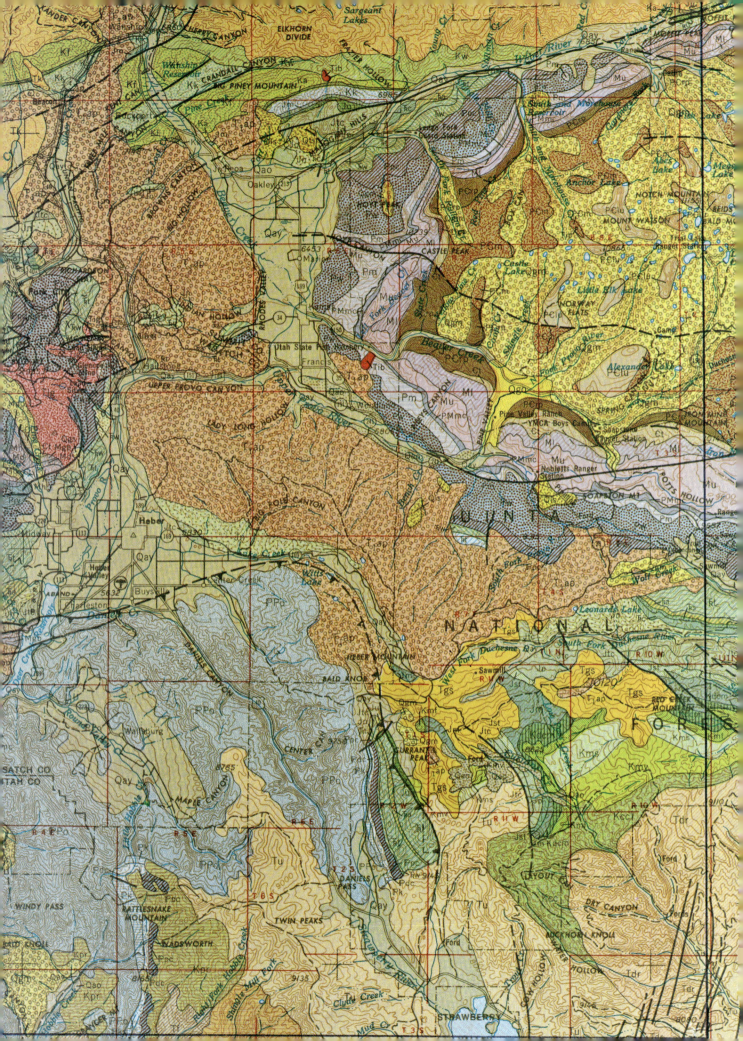

Canadian Rockies

Lake Minnewanka Quad. (West Half)
Alberta, Canada

Scale: 1.25 inch = 1 mile; 2 cm = 1 km

1. Generally characterize the rocks shown on this map as to age and environment.

2. Characterize the structure of the area as gently, moderately, or complexly deformed. What types of structures are present?

3. In which direction did the thrust faults move? How does this fit the plate tectonic model?

5. What is the likely age of deformation? What is the evidence for your answer?

4. Considering the plate tectonic model, what type of deformation is present?

6. What orogeny is represented by this tectonism?

LAKE MINNEWANKA

QUATERNARY

Qd	Light yellow	Till, alluvium, and colluvium

CRETACEOUS

Kbz	Dark green	Brazeau Formation Sandstone, siltstone, mudstone, and minor coal

JURASSIC

Jf	Dark blue green	Fernie Group Shale, siltstone, sandstone, and limestone

TRIASSIC

Ƭɼwh	Light green	Whitehorse Formation Sandstone, siltstone, and limestone breccia
Ƭɼrsm	Light yellow green	Sulfur Mountain Formation Siltstone, mudstone, and shale

PERMIAN AND PENNSYLVANIAN

PlPrm	Medium blue	Rocky Mountain Group Sandstone, dolomite, and chert

MISSISSIPPIAN

Met	Light gray	Etherington Formation Limestone and dolomite, cherty
Mmh	Light blue	Mount Head Formation Limestone and dolomite, cherty
Mtv	Dark gray	Turner Valley Formation Limestone and cherty limestone
Msh	Dark pinkish gray	Shunda Formation Limestone and cherty limestone
Mpk	Grayish pink	Pekisko Formation Limestone and cherty limestone
Mlv	Medium pinkish gray	Livingstone Formation Limestone, dolomite, and cherty limestone
Mbfu	Light pinkish gray	Upper part Exshaw and Banff Formations Limestone and dolomite
Mbfm	Medium pink	Middle part Exshaw and Banff Formations Limestone and dolomite
Mbfl	Purple	Lower part Exshaw and Banff Formations Shale, siltstone, and cherty limestone
Mbf	Light pink	Exshaw and Banff Formations undivided Limestone, shale, siltstone, cherty limestone, and dolomite

DEVONIAN

Dpa	Medium blue	Palliser Formation Dolomite and limestone
Dax	Dark blue	Alexo Formation Breccia
Dsx	Brown	Southesk Formation Dolomite
Dcn	Bluish gray	Cairn Formation Dolomite and limestone

CAMBRIAN

€lx	Bright blue	Lynx Group
		Dolomite and shale
€ar	Medium orange	Arctomys Formation
		Siltstone and dolomite
€pk	Light orange	Pika Formation
		Limestone, dolomite, and minor shale
€el	Medium gray	Eldon Formation
		Limestone and dolomite
€st	Dark yellow	Stephen Formation
		Shale and limestone
€ca	Light bluish gray	Cathedral Formation
		Limestone
€mw	Medium grayish pink	Mount Whyte Formation
		Shale, limestone, and dolomite

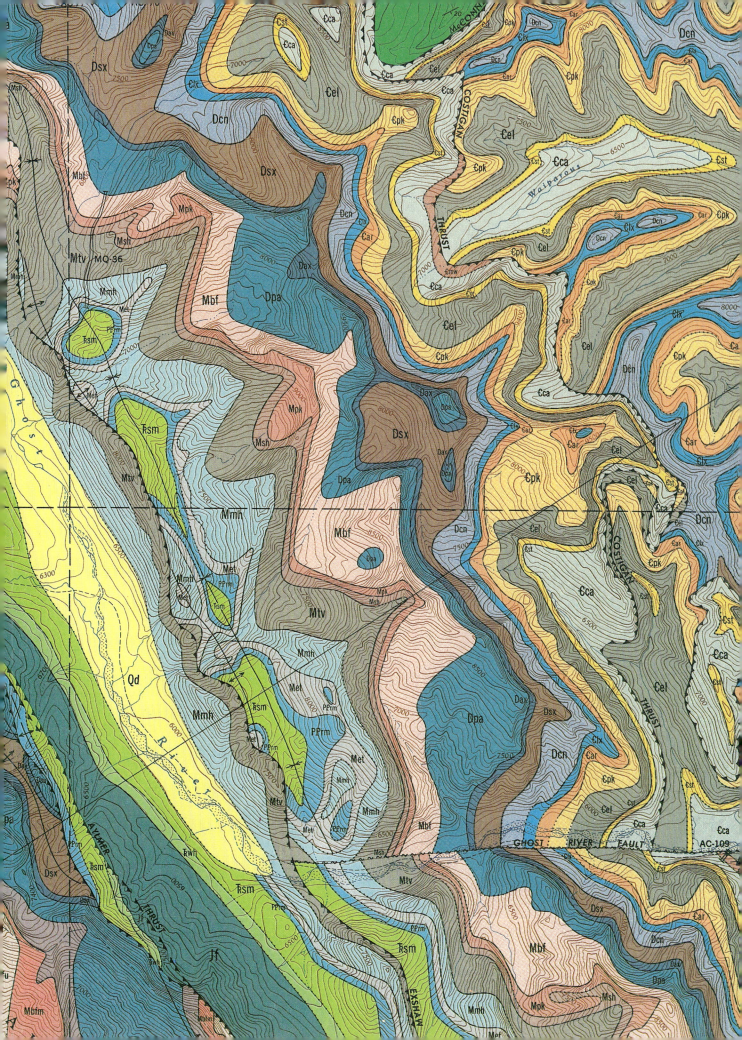

Black Hills and the High Plains

South Dakota State Geologic Map

Scale: 1 inch = 8 miles

1. What are the ages of the rocks shown on the map?

2. What type of geologic structure is represented by the belted elliptical outcrops?

3. What kind of rocks are exposed in the core of the Black Hills, along the flank, and in the plains to the east?

4. What is the age of the uplift that produced the Black Hills?

5. Locate and describe a nonconformity, disconformity, and two angular unconformities on this map.

6. Where would be the most likely area in the Black Hills for commercial mineral production?

SOUTH DAKOTA STATE GEOLOGIC MAP

TERTIARY

Tw	Light yellow	White River Group

CRETACEOUS

Kp	Light green, horizontal	Pierre Shale
Kn	Light green, vertical	Niobrara Formation
Kc	Light green, solid	Carlile Shale
Kg	Light green, solid	Greenhorn Limestone
Kbm	Light green, horizontal	Belle Fourche and Mowry Shales
Ksi	Green, fine vertical	Skull Creek Shale, Inyan Kara Group

JURASSIC

Jm	Grayish green	Morrison Formation
Js	Grayish green, horizontal	Sundance Formation

TRIASSIC

₸s	Blue green, diagonal	Spearfish Formation

PERMIAN

Pm	Blue, solid	Minnekahta and Opeche Formations

PENNSYLVANIAN

Cm	Light blue, fine horizontal	Minnelusa Sandstone

MISSISSIPPIAN

Cp	Light blue, diagonal	Pahasapa Limestone and Englewood Limestone

ORDOVICIAN

Ow	Purple	Whitewood Limestone

CAMBRIAN

€d	Orange brown	Deadwood Formation

PRECAMBRIAN

p €g	Dark brown	Granite and pegmatite
p €a	White, brown horizontal	Basic igneous intrusions
p €sq	Light brown, horizontal, white irregular dashes	Schist and quartz
p €sc	Light brown, diagonal	Sandstone and conglomerate

IGNEOUS ROCKS

Qr	Bright green	Rhyolite and obsidian
Tr	Red and white	Rhyolite
Tp	Red	Light-colored intrusive igneous rocks

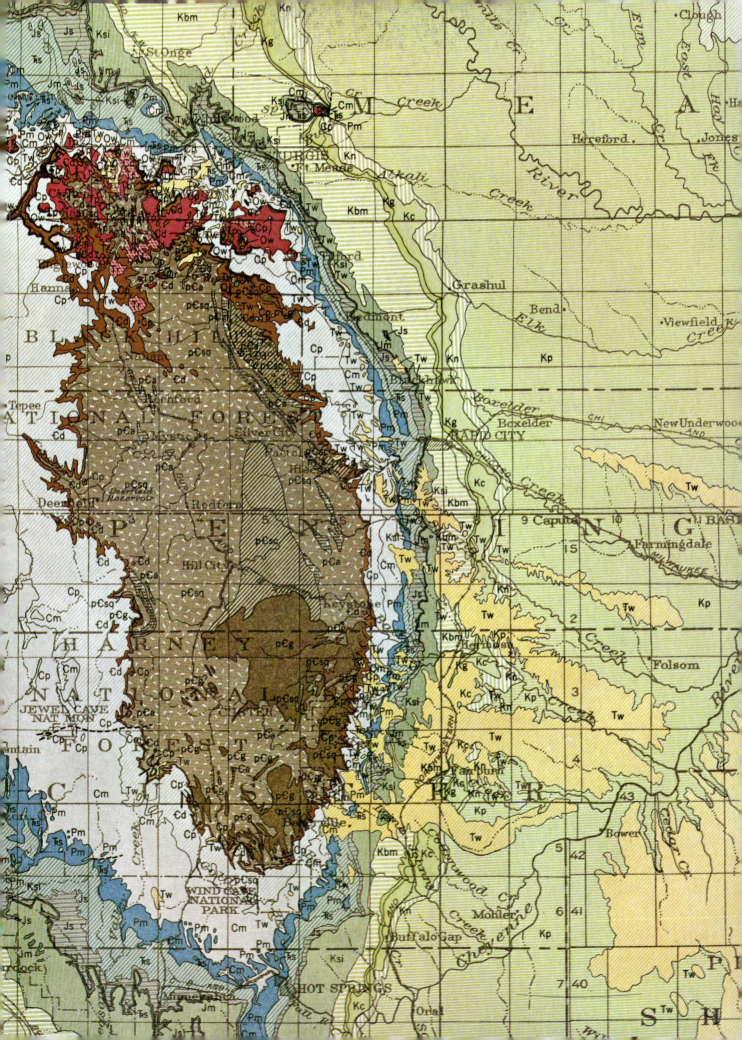

Sierra Nevada

Geology of Mono Craters Quad. California

Scale: 1 inch = 1 mile

1. What are the ages and general rock types exposed in this area?

2. Determine a generalized sequence of events for the history of this area.

3. This area is part of the Sierra Nevada batholith complex. Draw a contact between the batholith and the country rock. What evidence of multiple intrusion in the Sierra Nevada is shown on the map?

6. What is the origin of the lobate deposits in the Lee Vining, Walker Lake, and Gull Lake areas?

4. What evidence is there of block-faulting in the area? What are the ages of these faults?

7. Within the plate tectonic model, what type of plate interaction would explain the large quantity of intrusive igneous rocks present in the mapped area?

5. Study the Quaternary sediments. What evidence of glaciation is present in this area?

SIERRA NEVADA MONO
CRATERS QUADRANGLE

ROOF PENDANT SEQUENCES

JURASSIC

Jc	Light blue green, solid	Conglomerate
Jt	Light green, solid	Volcanic tuffs, flows, and shale
Js	Greenish gray, solid	Marble and metamorphosed shale
Jx	Light green, solid	Volcanic rocks and cross-bedded sandstone

PENNSYLVANIAN-PERMIAN

Ph	Light blue, solid	Contact metamorphic rocks and marble
Pa	Bright green, solid	Conglomerates
Pt	Medium blue gray, solid	Volcanic rocks and dirty sandstones
Pc	Light yellow green, solid	Andesite and sandstone
PIPh	Light brownish gray, solid	Contact metamorphic rocks, volcanic flow
PIPq	Medium purple brown, solid	Quartzite, hornfels

SILURIAN-ORDOVICIAN

SOs	Red stippled grayish lavender	Marble and contact metamorphics
SOh	Dark lavender	Contact metamorphics
SOc	Bluish gray	Quartz, marble, contact metamorphics
SOa	Pinkish gray	Contact metamorphics, thin marbles
SOq	Light grayish brown	Quartzite
SOm	Grayish pink	Marble, contact metamorphics

INTRUSIVE IGNEOUS ROCKS

TERTIARY

Tda	Red, orange	Shallow intrusives

CRETACEOUS

Ka	Light tan, solid	Quartz monzonite
Kja	Lavender, solid	Granite
Kk	Grayish lavender, solid	Granodiorite
Kgu	White, pink irregular dashes	Granite

JURASSIC

Jd	Light purple, solid	Diorite
Jl	Light orange, solid	Granodiorite
Jla	Light orange with red crosses	Garnet-bearing, light-colored intrusives
Jm	Dark pink, solid	Monzonite

CENOZOIC SEDIMENTS

QUATERNARY

Qm	Yellow, red stippled	Cirque moraine
Qrg	Yellow, black triangles	Rock glacier
Qsl	Yellow, black circles	Rock glacier
Qal	Yellow, black stippled	Alluvium and pumice
Qts	Light yellow	Talus and slopewash
Qti	Light yellow, green stippled	Till of Tiago Glaciation
Qta	Yellow, brown circles	Till of Tahoe Glaciation
Qtao	Stippled green and yellow	Till of Tahoe Glaciation
Qsh	Light yellow, green stippled	Till of Sherwin Glaciation
Qto	Light yellow, green stippled	Till of Sherwin Glaciation

TERTIARY

Tcl	Tan	Volcanic conglomerates

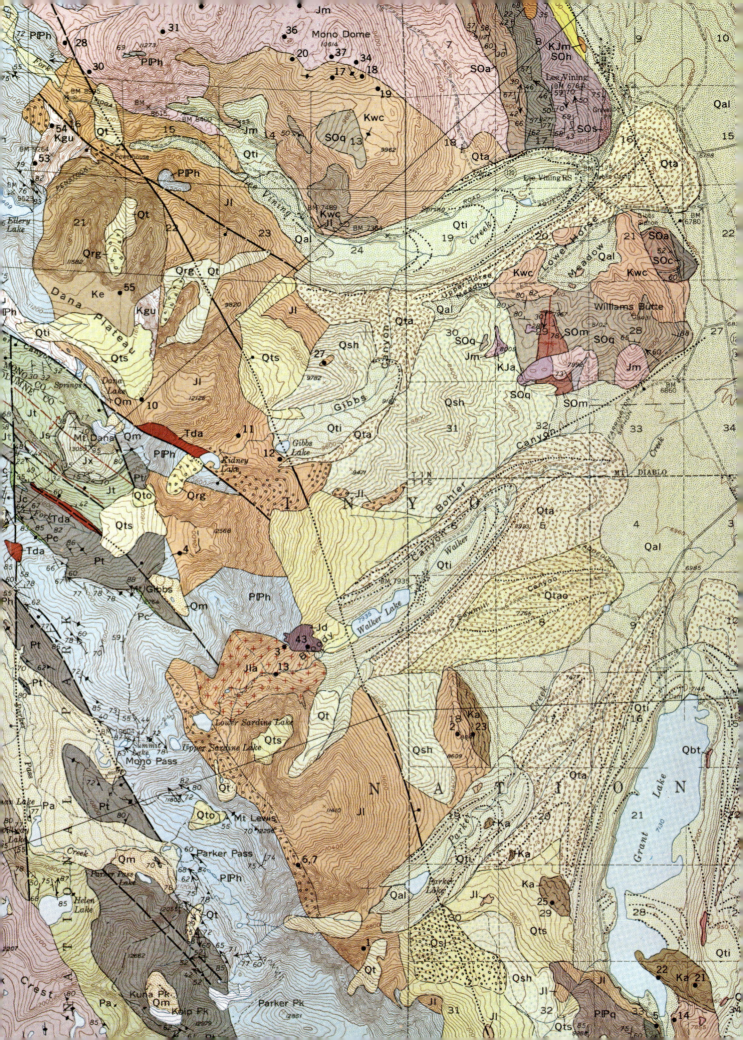

Block-Fault
Mountains of the
Basin and Range

Geologic Map of Clark County, Nevada

Scale: 1 inch = 4 miles

1. What are the ages of the rocks in the map area? Is this typical for the entire Basin and Range area?

2. What type of faulting is present in the north end of the Sheep Range approximately 5 miles northwest of Mormon Well?

3. What type of faulting is present in the area approximately 5 miles south and east of Fossil Ridge in the east-central part of the map?

4. What is the age of the major thrust fault along the east side of Wilson Cliffs, approximately 20 miles due west of Las Vegas?

5. Explain the alternating valley and mountain range pattern as present along the northern margin of the map area.

6. What is the explanation for the northwest trend of Las Vegas Valley, and the lack of continuity of mountain ranges on either side of the valley?

7. Are these rocks deformed by tension, compression, or shear? Explain your answer in terms of plate tectonics.

CLARK COUNTY, NEVADA

QUATERNARY

| Qal | Light yellow | Alluvium |
| Ql | Bright yellow | Las Vegas Formation |

TERTIARY

| Th | Red on yellow diagonal | Horse Spring Formation |

JURASSIC

| Ja | Medium blue green | Cross-bedded sandstone |

TRIASSIC

| Ʀcm | Light green | Chinle and Moenkopi Formations |
| | | Red siltstone and shale |

PERMIAN

| Pkt | Blue | Kaibab, Toroweap, and Coconino Formations |
| | | Limestone, siltstone, and sandstone |

PENNSYLVANIAN

| PlPMb | Light blue | Bird Spring Formation |
| | | Thin-bedded cherty limestone |

MISSISSIPPIAN

| Mm | Blue and purple horizontal | Monte Cristo |
| | | Thick-bedded cherty limestone |

DEVONIAN

| Ds | Lavender | Sultan Limestone |

SILURIAN

| Sl | Purple | Lone Mountain Dolomite |
| | | Medium gray crystalline dolomite |

ORDOVICIAN

Oep	Reddish purple,	Ely Springs Dolomite, Eureka Quartzite, and
	with red marker horizon	Pogonip Group
		Dark gray dolomites, light gray quartzite, thin-bedded
		fossiliferous limestone
		Eureka Quartzite is the red marker horizon

CAMBRIAN

| €dl | Brown and tan diagonal | Undivided Middle and Upper Cambrian limestone and dolomite |
| €u | Green and brown diagonal | Undivided Lower and Middle Cambrian quartzite, shale, and limestone |

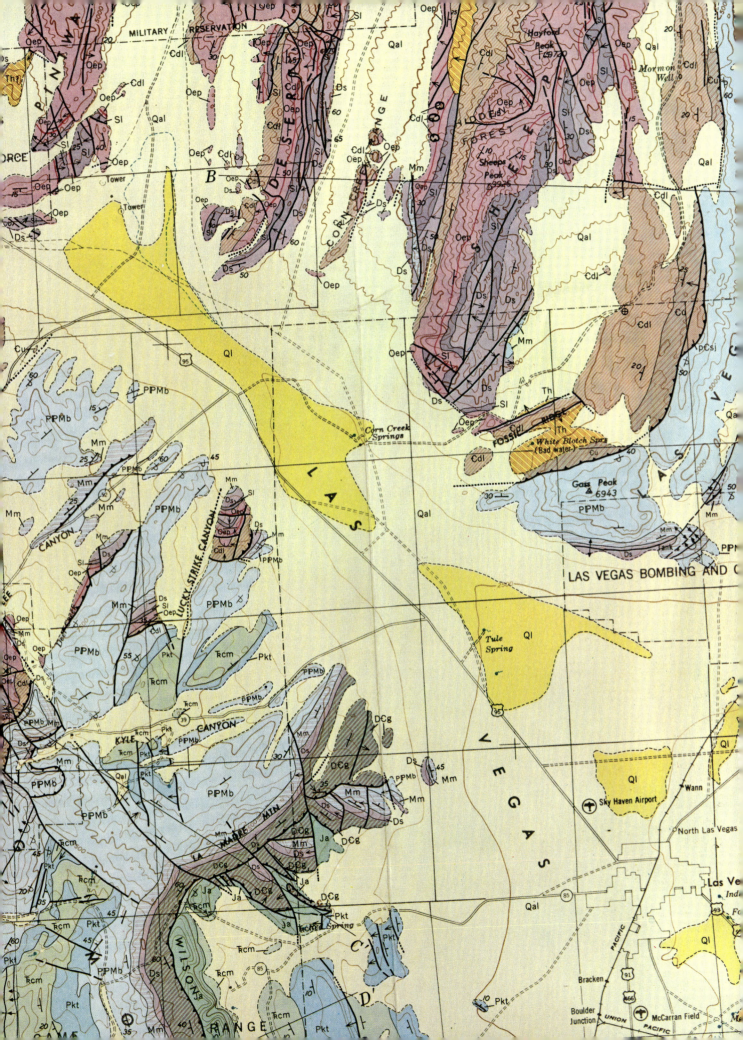

Active Coast Ranges of California

Geology of Valyermo Quad. California

Scale: 1 inch = 2000 feet, or approximately 0.4 mile

1. The San Andreas Fault is a classic example of a trans-form fault. What is the age of movement along the San Andreas Fault zone? What is your evidence from the map?

2. What is the youngest folded rock unit in this area?

3. What orogenic episode would the deformation in this area represent?

4. Trace the normal or unfaulted contact between the granodiorite and the Martinez Formation on Pinyon Ridge. What is the nature of the contact?

5. What evidence of continental drift is presented in this area?

6. What type of plate interaction explains the most recent deformation of the rocks in this area?

VALYERMO QUADRANGLE

QUATERNARY

Qya	Yellow, gold overlay	Younger alluvium
Qoa	White, orange stipples	Older alluvium
Qs	Gold, light brown dots	Shoemaker gravel, gravel, and interbedded sand and silt
Qh	Orangish brown, crossed lines	Harold Formation
		Siltstone with interbeds of conglomerate and shale

TERTIARY

Tpw	Green, fine diagonal	Punchbowl Formation
		Shale, sandstone, and conglomerate northeast of San Andreas Fault
Tpu	Fine purple, diagonal	
Tpl	Fine blue, diagonal	Upper and Lower Punchbowl Formation
		Southwest of San Andreas Fault, well-consolidated conglomeratic sandstone
Tmu	Orange	Upper and Lower Martinez Formation
Tml	Yellow	Interbedded sandstone, shale, and conglomerate

CRETACEOUS

Khb	Dark bluish green	Crushed basement rocks near San Andreas Fault
Kh	Turquoise, diagonal	Holcomb quartz monzonite

JURASSIC

Jp	Gray and blue diagonal	Granodiorite

PALEOZOIC AND MESOZOIC

Crush Zone	Dark red and white diamonds	Crushed basement rocks
Pv	Red and white, diagonal	Intrusive and metamorphic rocks

PALEOZOIC

Pl	Pink and red, vertical	Limestone

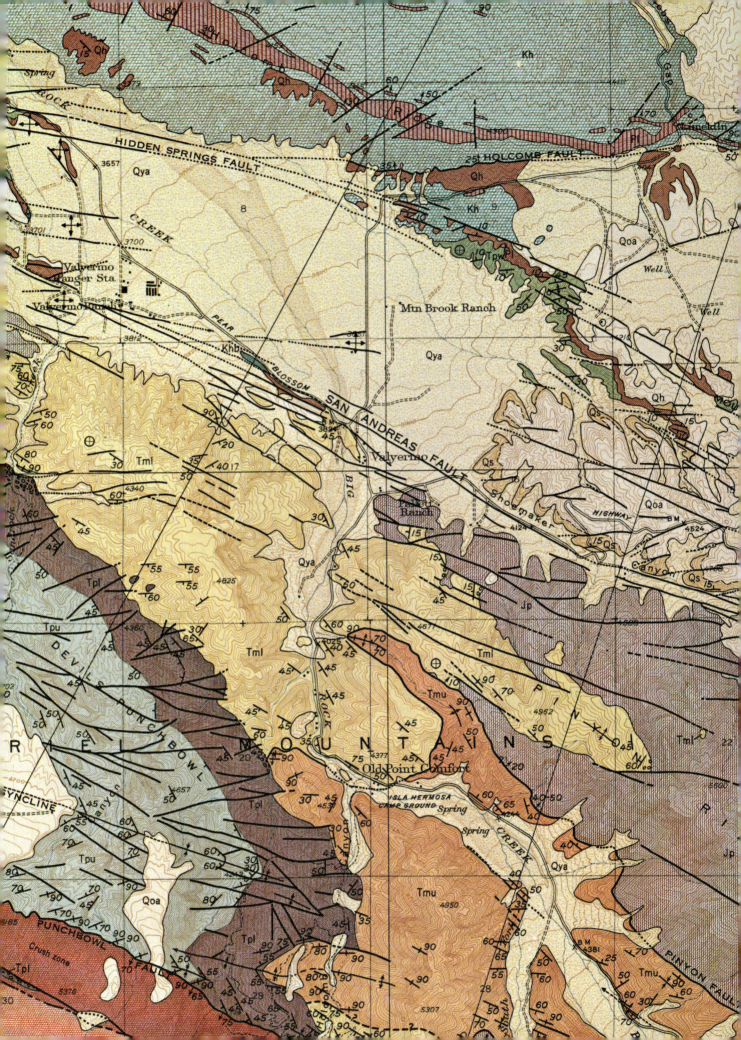

Arctic Islands

Geologic Map of North America

Scale: 1 inch = approximately 80 miles

1. What are the ages of the rocks shown on the map?

2. What is the pattern of the distribution of the different ages of rock in the map area?

3. What is the significance of the lack of Cambrian and Mississippian rocks in this area?

4. What is the age of the folding on the Parry Islands in the west-central part of the map?

5. What is the evidence for continental accretion in this area?

6. The basic structure of North America is nearly symmetrical with a central stable area rimmed by mobile belts. The Appalachian mobile belt occurs on the east and south, and the Cordilleran mobile belt occurs on the west of the stable interior. What evidence is there in the map area of a northern mobile belt?

7. Is there any evidence of convergent plate interaction in this area? Explain your answer.

ARCTIC ISLANDS

QT	Gold and orange, diagonal	Quaternary and Upper Tertiary
T	Yellow gold	Tertiary
Tc	Yellow	Tertiary continental
Mz	Light green, horizontal	Mesozoic sedimentary
Ku	Brownish green	Upper Cretaceous
Kl	Olive green	Lower Cretaceous
KJ	Green	Cretaceous and uppermost Jurassic
Tr	Light green	Triassic
Pz	Light purple	Paleozoic
P	Light blue	Permian
PC	Blue	Permian and Carboniferous
IP	Bright blue	Pennsylvanian
D	Brownish gray	Devonian
D_2	Brown	Upper Devonian
D_1	Bluish gray	Lower Devonian
DSO	Blue and pink, diagonal	Devonian, Silurian, and Ordovician
SO	Pinkish purple	Silurian and Ordovician
S	Light purple	Silurian
O	Reddish purple	Ordovician
p€u	Tan	Upper Precambrian
P€i	Pink	Precambrian granite and granite gneiss
p€u_2	Reddish brown	Keweenawan sedimentary
p€u_1	Brown	Keweenawan volcanic
p€m	Greenish brown	Middle Precambrian

31

Pleistocene Glaciation

Glacial Map of North America

Scale: 1 inch = 72 miles

1. On the basis of the principle of superposition, what are the relative ages of the four main Pleistocene till sheets?

2. What was the direction of ice movement in the vicinity of Chicago, Illinois; Fort Wayne, Indiana; and Mankato, Minnesota?

3. What were the likely controlling factors that determined the present position of the Missouri and Ohio rivers?

4. Explain the lack of glacial deposits in the southwestern corner of Wisconsin.

5. Pleistocene lake deposits are illustrated on the map by a horizontal pattern. What is the origin of the Pleistocene lake basin approximately 50 miles south of Chicago?

6. What type of isostatic forces are likely to be affecting this area at the present time? Explain your answer.

PLEISTOCENE

Pleistocene Lakes	Blue horizontal	
Wisconsin Glaciation	Dark pink	Moraine
	Light pink	Intermoraine
Illinoian Glaciation	Dark green	Moraine
	Light green	Intermoraine
Kansan Glaciation	Light yellow	Glacial deposits
Nebraskan Glaciation	Yellow, vertical	Glacial deposits

Hominid Fossils

The first significant hominid fossils were found north of Düsseldorf, Germany, in the Neander Valley in 1856. From then until now numerous finds have expanded the fossil collections of the Family Hominidae. These fossil remains are among the most valuable objects of antiquity in the possession of humankind. They are literally "national treasures" housed in approximately 25 museums around the world where they are given the utmost protection in safety and environmental controls. There are so few existing specimens that each one can have a significant effect on the theory of hominid development. Examples of single finds that have revolutionized our evolutionary theories include the discovery of the "Taung baby" in Botswana in 1924, *Homo habilis* in Tanzania in 1962, "Lucy" in Ethiopia in 1974, 3.6 to 3.8 million year old footprints ("first family") in Laetoli in Tanzania in 1976, the finding of the "black skull" in Kenya in 1985, and "Turkana boy" in Kenya in 1984. Each discovery represents either a profound bit of luck by the finder or, in most cases, the result of very expensive, painstaking, and time-consuming excavations whose purpose was to seek hominid fossils in known productive locations. As a collection, they represent the evolutionary record of humankind for approximately the past 5 million years.

The classification of the hominids is in a state of flux. Although there is almost no disagreement among scientists about the fact of hominid evolution, uncertainty exists over the specifics of evolutionary relationships. There is widespread belief that science teaches that humans were derived from the apes. This is not so. Modern hominids did not evolve from modern pongids; both are contemporaneous.

They might, or might not, have shared a common ancestral stock. Whether they both evolved from a single common family, or each followed independent lines of evolution, is not clear. The hominid fossils presently available to scholars provide us with a reasonably good look at the evolutionary history of the family. The evolutionary sequence we see from these specimens elucidates hominid origins at a time that was perhaps close to the probable time of divergence, if indeed hominids and pongids do share a common ancestor. The missing information at present is the dearth of fossil pongids over the past 5 million years. If we had access to comparable pongid fossils covering this interval, the evolutionary story of these two families would be much more clear, and the identity of a possible common ancestor would be more predictable.

Most taxonomic schemes recognize two genera of hominids: *Australopithecus* and *Homo.* Within *Australopithecus,* six species have been differentiated: *Australopithecus afarensis, Australopithecus africanus, Australopithecus anamensis, Australopithecus aethiopicus, Australopithecus boisei,* and *Australopithecus robustus.* Seven species are recognized in the genus *Homo: Homo rudolfensis, Homo habilis, Homo ergaster, Homo erectus, Homo heidelbergensis, Homo neanderthalensis,* and *Homo sapiens.* A third genus of the Hominidae has been described and named: *Ardipithecus,* with a single species: *ramidus.* The material for this taxon was found in Ethiopia in 1994–1995. Table 32–1 presents a tabular summary of the above listed taxa. Figure 32.1 illustrates the general evolutionary relationships thought to exist among hominid species.

Table 32–1 Taxonomic Scheme of Hominids

Name	Age (m.y.)	Sites	Special Features
Homo sapiens	0.1–0.0	Worldwide	Oldest cave paintings, c. 31,000 years, includes Cro-magnon.
Homo neanderthalensis	0.3–0.03	Europe, Asia, Africa	Discovered 1856 in Neander Valley in northern Germany. Most productive site in Croatia, where 850 fossils have been found.
Homo hiedelbergensis	0.6–0.2	Europe	Discovered in 1921, first fossil hominid found in Africa. First to build shelters.
Homo erectus	1.2–0.4	North and East Africa, Indonesia, China	First to use controlled fire. Includes "Peking Man" and "Java Man."
Homo ergaster	1.8–1.5	East Africa	"Turkana boy," oldest complete skeleton found, stood nearly 6 feet tall.
Homo habilis	1.9–1.6	East Africa	Smallest adult hominid ever found, height estimated to be approximately 1 meter.
Homo rudolfensis	2.4–1.9	East Africa	Similar to *H. habilis.* The two lived comtemporaneously in East Africa.
Homo sp.	2.5–?	East Africa	A single upper jaw (maxilla) collected from a site otherwise restricted to specimens of *Australopithecus.* Stone tools present.
Australopithecus robustus	1.9–1.0	South Africa	Originally called *Paranthropus robustus,* and many workers prefer the older term.
Australopithecus boisei	2.3–1.4	East Africa	Became famous as *Zinjanthropus,* the nutcracker man, a hyper-robust species.
Australopithecus aethiopicus	2.7–1.9	East Africa	"Black skull," robust australopithecine with a prominent sagittal crest and huge teeth.
Australopithecus africanus	2.8–2.4	South Africa	"Taung child" found in 1924, highly controversial for many years, a fossil gem.
Australopithecus afarensis	3.9–3.0	East Africa, Central Africa	"Lucy" and the Laetoli footprints. Includes *A. bahrelghazali* found in central Africa.
Australopithecus anamensis	4.2–3.9	East Africa	Material exhibits an unusual combination of *Homo* and *Australopithecus* characters.
Ardipithecus ramidus	4.4	East Africa	A third genus within the Hominidae, thought to be ancestral to *Australopithecus.*

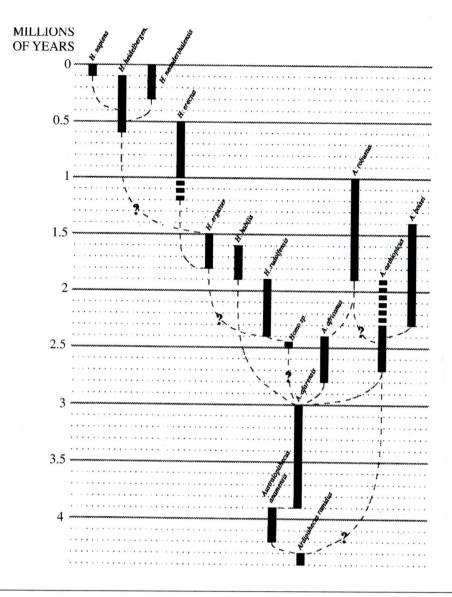

Figure 32.1 Diagram showing temporal and evolutionary relationships among fossil hominids. (Reprinted with permission from *Nature*, Bernard A. Wood, The Oldest Hominid Yet, 371:280-281, 1994, Macmillian Magazines Limited.)

Procedure

1. Figure 32.2 illustrates a comparison of the main characteristics of skull morphology of a chimpanzee, representing the apes (Family Pongidae), with a modern human skull (Family Hominidae). Study the important differences between these two skulls.

2. Examine the drawings (fig. 32.3) of the six fossil skulls noting the similarities with the chimp and modern human skulls. The natural stratigraphic order of these skulls has intentionally been altered for your later reconstruction.

3. Reconstruct what you think to be the natural stratigraphic sequence of the fossil skulls based upon their evolutionary development. Label the individual skulls from 1 to 6 in order of their similarity to the chimp and human skulls (1 being most apelike and 6 being most humanlike).

4. Compare your evolutionary sequence with that of your instructor.

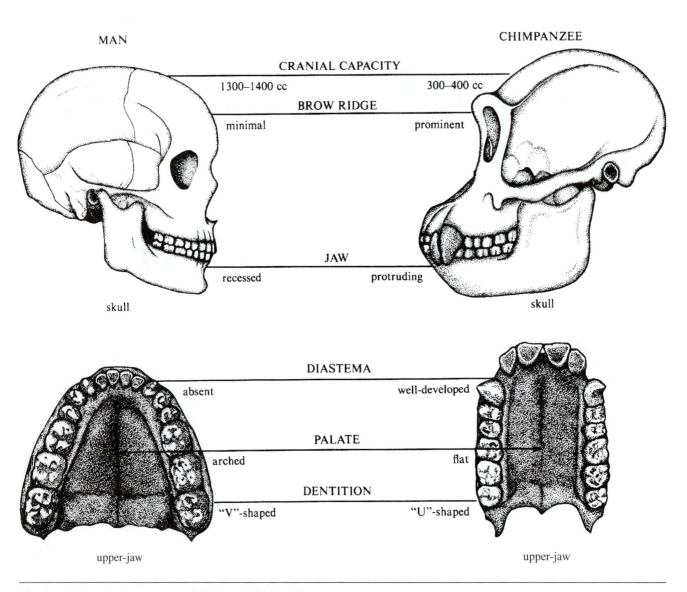

Figure 32.2 Characteristics of skull morphology for a human and a chimp skull and upper jaws.

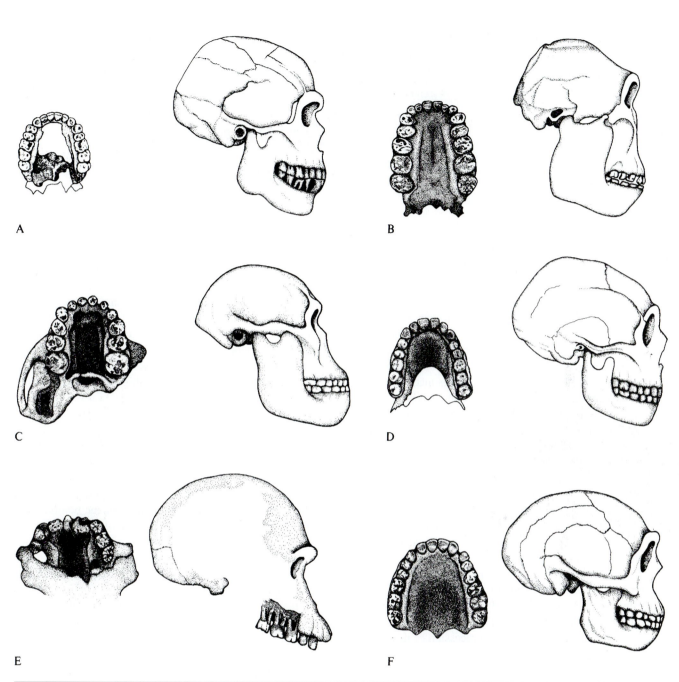

Figure 32.3 Selection of fossil hominid skulls and their dentition. The stratigraphic order of their occurrence has intentionally been altered awaiting the student's evaluation. The jaw associated with each skull is located to its left.

The Expanding Universe

In order to place the age of the Earth in its proper perspective, it is useful to understand the modern theories concerning the origin of the universe and our own Milky Way galaxy of which our solar system is a part. In the early 1930s, two American astronomers, Humason and Hubble of the Mount Wilson Observatory in California, studied the characteristics and spacing of spectral lines made on photographic plates of light emanating from distant galaxies. Their discovery that the spectral lines from these galaxies shifted toward the red end of the spectrum led to the **red shift theory.** As Humason (who, by the way, came to the Mount Wilson Observatory as a mule skinner with an eighth-grade education and, like Carl Sagan, died rich in the respect of his colleagues) and Hubble reasoned, this unusual shift could best be explained as similar to the Doppler effect in sound transmission. As the pitch of a horn or whistle on a moving vehicle changes to a lower frequency when a vehicle passes an observer, so the light from the galaxies is changed by being shifted toward the red, or longer wavelength, as the galaxy moves away from Earth.

Furthermore, these men demonstrated a proportional relationship in the amount of red shift and the velocity of the receding galaxy. The faster the recession of the galaxy, the greater the amount of red shift.

From the work of Humason and Hubble has grown the **theory of the expanding universe,** which states that 10 to 20 billion years ago all matter and energy in the universe were compacted in a single place having a very small volume. This extremely dense accumulation of matter and energy exploded sending material outward in all directions. Following this "big bang" all the galaxies that exist today were formed as material accumulated around specific rotating centers. The red shift in the spectral patterns of light emanating from galaxies is the result of their speeding outward in an expanding pattern that drives them further and further apart.

By plotting the radial velocities of several galaxies in kilometers per second against the distances from the Earth of the same galaxies, it is possible to calculate the time since the "big bang," which is the age of the universe as we know it. The slope of a line formed by the intersection of velocity and distance values for several galaxies allows us to project backward in time to the "big bang."

Procedure

On the graph provided in figure 33.1, plot radial velocity and distance data for the five galaxies given in table 33–1. (Plot the velocity on the horizontal axis and the distance on the vertical axis.) Using your five points as a guide, draw a straight line to best fit the distribution of the points. Make sure your line passes through the origin of the graph.

Using any two sets of values, calculate the slope of the line you have constructed. What is the time in years (convert seconds to years) since the "big bang," based upon your calculation?

Figure 33.2 illustrates the five galaxies in table 33–1.

Table 33–1 Radial Velocity (velocity of recession from Earth) and the Distance from Earth of Five Galaxies Whose Measurements Have Been Determined by Astronomers

Galaxy	Radial Velocity (km/sec)	Distance (km)
Virgo	1.2×10^3	78×10^{19}
Ursa Major	15×10^3	1000×10^{19}
Corona Borealis	22×10^3	1400×10^{19}
Bootes	39×10^3	2500×10^{19}
Hydra	61×10^3	3960×10^{19}

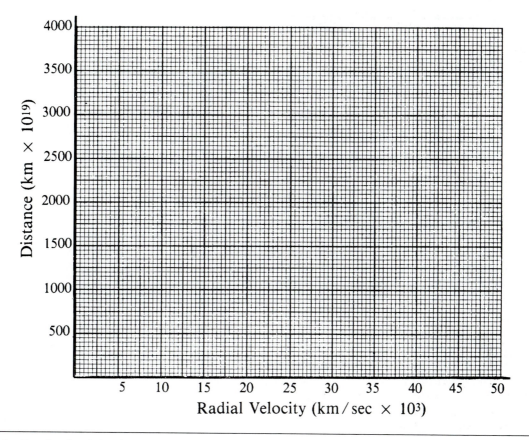

Figure 33.1   Graph of radial velocity and distance for plotting data from table 33–1 for use in determining the age of the universe.

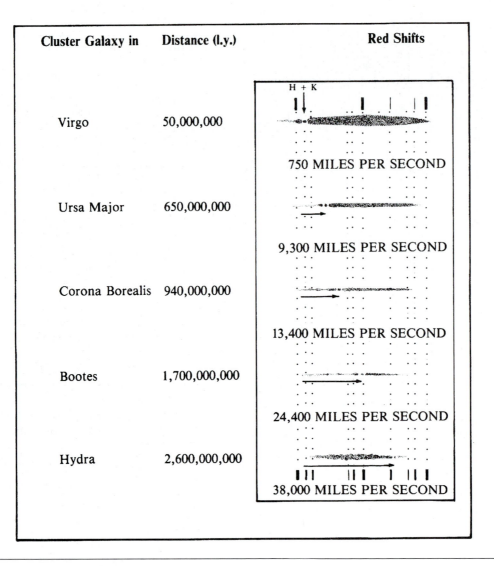

Cluster Galaxy in	Distance (l.y.)	Red Shifts
Virgo	50,000,000	750 MILES PER SECOND
Ursa Major	650,000,000	9,300 MILES PER SECOND
Corona Borealis	940,000,000	13,400 MILES PER SECOND
Bootes	1,700,000,000	24,400 MILES PER SECOND
Hydra	2,600,000,000	38,000 MILES PER SECOND

Figure 33.2 Illustration showing the five galaxies used in table 33–1 and the Red Shift of the H and K lines of ionized calcium. The arrows show the amount of shift for these lines on each of the five galaxies. The actual lines are shown by the two breaks against the disk of the galaxy. (After Abell, G., 1964. *Exploration of the Universe,* Holt, Rinehart & Winston, p. 580.)

Review of Topographic Map Reading

Contour lines express points of equal elevation above sea level. Remarkable accuracy as well as ease of reading make this method of modeling truly ingenius. All subtleties of topography are clearly illustrated by the patterns of contour lines.

Contours are drawn every few feet depending on the ruggedness of the topography. In flat lands, contour lines expressing changes in elevation of 5 feet are ideal, whereas, in mountainous terrain, intervals of 50, 100, or 200 feet are necessary to keep the patterns clear and easy to read. Every fifth line is drawn as a heavy contour, expressing larger increments for rapid determinations. These heavy contours are also marked with the elevation. The reader must count up or down from the heavy or index contours, applying the contour interval printed on the map, to determine intermediate elevations.

Figure A.1 illustrates the application of contouring. Simple rules to remember in reading contour maps are:

1. Hills or peaks are expressed as closed patterns.
2. Depressions are closed patterns with inward pointing barbs on the line.
3. Contour lines "V" upstream as they cross valleys.
4. The closer the contour lines the steeper the elevation.
5. Contour lines never cross or divide.

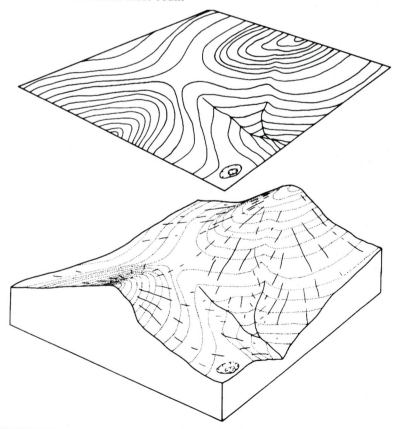

Figure A.1 Contouring.